CALIFORNIA NATURAL HISTORY GUIDES

INTRODUCTION TO CALIFORNIA SPRING WILDFLOWERS

California Natural History Guides

Phyllis M. Faber and Bruce M. Pavlik, General Editors

Introduction to CALIFORNIA SPRING WILDFLOWERS

REVISED EDITION

Philip A. Munz

Edited by Dianne Lake and Phyllis M. Faber

UNIVERSITY OF CALIFORNIA PRESS

Berkeley Los Angeles London

California Natural History Guides No. 75

University of California Press
Berkeley and Los Angeles, California

University of California Press, Ltd.
London, England

Library of Congress Cataloging-in-Publication Data

Munz, Philip A. (Philip Alexander), 1892–
Introduction to California spring wildflowers of the foothills, valleys and coast / Philip A. Munz ; edited by Dianne Lake and Phyllis M. Faber.—Rev. ed.
p. cm.
Rev. ed. of: California spring wildflowers, from the base of the Sierra Nevada and Southern Mountains to the sea. 1961.
ISBN 0-520-23633-5 (case)—ISBN 0-520-23634-3 (paper)
1. Wild flowers—California—Identification. 2. Wild flowers—California—Pictorial works. I. Lake, Dianne. II. Faber, Phyllis M. III. Munz, Philip A. (Philip Alexander), 1892– California spring wildflowers, from the base of the Sierra Nevada and Southern Mountains to the sea. IV. Title.

QK149.M794 2004
582.13′09794—dc22 2003055851

Manufactured in China

10 09 08 07 06 05 04
10 9 8 7 6 5 4 3 2 1

The paper used in this publication meets the minimum requirements of ANSI/NISO Z39.48–1992 (R 1997) (*Permanence of Paper*). ♾

Cover: Shooting star, Mission Trails, Regional Park, San Diego.
Photograph by Christopher Talbot Frank.

The publisher gratefully acknowledges the generous contributions to this book provided by

the Gordon and Betty Moore Fund
in Environmental Studies
and
the General Endowment Fund of the
University of California Press Associates.

Grateful acknowledgment is also made to
William T. and Wilma Follette
and
to the California Academy of Sciences

for their contribution of photographs.

CONTENTS

EDITOR'S PREFACE TO THE NEW EDITION

This new edition of Philip A. Munz's *California Spring Wildflowers* has been extensively revised to make the book more useable statewide. Our goal was to retain Munz's intent but to add more descriptive material to help identify plants. Many rare plants and less showy ones were replaced with species that are more common statewide and more readily seen. Plant names for each plant have been updated to conform to the current authority, *The Jepson Manual: Higher Plants of California,* J. Hickman, editor (University of California Press, 1993). Each plant name has been checked for accuracy and currency. In several cases, research done in the last 50 years has resulted in some species being absorbed into other species, and some being split into varieties or subspecies. In addition, some varieties and subspecies have even become separate species. All these changes have been incorporated into the text.

Each plant has been given a common name using the following sources listed here in descending order of preference: the *Jepson Manual;* Philip Munz, *California Flora* (University of California Press, 1959); and Leroy Abrams, *Illustrated Flora of the Pacific States* (Stanford University Press, 1923–1960). The rule developed by Munz for hyphenation has been used for all common names: if a plant's common name indicates a different genus or family, a hyphen is inserted to show that the plant does not actually belong to that genus or family. Thus, "skunk-cabbage" is hyphenated because the plant it refers to is not in the cabbage genus nor the cabbage family, but "tiger lily" is not hyphenated because the plant it refers to is in the lily genus, as well as in the lily family.

Geographical distribution and elevational information were also examined and changed where new data was available. A new geographical system was adapted for Munz's *Mountain Wildflowers* and *Spring Wildflowers* books because of the great number of changes that have resulted from exploration and research in those areas over the years. The new system was not used in the *Shore Wildflowers* book because the book deals only with the immediate coast and no other areas of California.

A great deal of new information has been gathered in the last 50 years. The ranges of many plants have expanded, and they have shrunk or changed for others because of human activity, taxonomic changes, or other factors. In the original text, Philip Munz used a combination of regions and counties to describe the range and distribution of plants, but because of the many changes resulting from the past 50 years of research, as well as the discovery of new populations almost every day, a purely regional system was adopted for the new edition since plants tend to adhere to regions determined by geography, geology, and climate more than they adhere to county lines. This purely regional system allows for more flexibility and tends to be more accurate and inclusive over time.

The regional distribution system adopted for this new edition is based on the one used in the *Jepson Manual*. The regional designations are the Sierra Nevada, the Coast Ranges (South, Central, and North), the Peninsular and Transverse Ranges, the Cascade Range, the Modoc Plateau, the Great Basin areas of eastern California, and the San Jacinto, San Bernardino, San Gabriel, Tehachapi, Warner, Sweetwater, White, Inyo, Klamath, and Desert Mountains. In addition, northwestern California is used to refer to the North Coast

Ranges and Klamath Mountains; and southwestern California refers to the Peninsular, Transverse, and South Coast Ranges.

One of the regional differences that should be noted between the original text and the new edition is in the use of the term "North Coast Ranges." Although Munz used this term to include both the coastal and central ranges of northern California, the new edition uses the *Jepson Manual* interpretation that defines the North Coast Ranges as consisting only of the inner and outer coastal mountains. The central, more inland ranges are referred to as the Klamath Mountains and include the Trinity, Marble, Scott, Salmon, and Siskyou Mountains.

Another regional description used here, but not in the original Munz text, is "the Great Basin areas of eastern California," which includes the Warner Mountains and Modoc Plateau, the White and Inyo Mountains, and the high desert areas east of the Sierra Nevada. Although Munz expressly states that he does not include any of the desert areas, this new edition does indicate if the range of a plant extends into the desert in addition to its purely montane distribution.

In the 1959 edition of this book, Munz included a number of showy trees throughout the book. These have been gathered into a separate section because they don't qualify as wildflowers but are showy and attractive and users of this book may wonder what they are.

Dianne Lake has revised this edition by selecting the plants to be included and writing new descriptions or revising original descriptions to make them readable, entertaining, and informative for today's readers. The Press is grateful to her for her careful and thorough work reflected in this book and for sharing her knowledge of the California flora. Drawings are from the original edition and were drawn by Dr. Stephen S. Tillett of the New York Botanical Garden, and a few were drawn by Richard J. Shaw of Utah State University. Color illustrations and new design features have been added to make the book more attractive and user-friendly.

The Press is particularly grateful to William T. Follette and his wife, Wilma, for so many of the photographs in this book. Beautiful photographs, reliably identified, are welcome additions to any botanical publication.

The late Dr. Robert Ornduff wrote an introduction for each of the four books in this series shortly before his untimely death in 2000.

This is the third of four books in the Munz wildflower series: *Shore Wildflowers* and *Mountain Wildflowers* both were published in 2003, and *Desert Wildflowers* will debut this year also.

Phyllis M. Faber
August 2003

INTRODUCTION

California has long been considered an earthly paradise, especially in spring when its rolling hills and green valleys are full of wildflowers. About 6,000 flowering plants occur in the state, many of which, like the grasses (Poaceae) and sedges (Cyperaceae), are very important for grazing but not of especial interest to the wildflower lover. Even when these plants, the trees, and the more inconspicuous bushes are deleted from the list, however, several thousand true wildflowers still remain.

When we recall the great variety of topographical conditions in California and the plants we see in its different areas, we know that the desert flowers are quite different from those on coastal slopes and that summer bloomers in high mountains differ from the spring plants of the valleys. Therefore, to bring before the general reader in compact and useful form something by which wildflowers can be identified, this introductory book is presented. The more discriminating student can turn to *The Jepson Manual: Higher Plants of California*, by J. Hickman (ed.) (University of California Press, 1993) for more detail.

Climatic Conditions

Between the mountains and the coast the topography exhibits considerable range. Some of it is wooded, some is brushy, and some is grassland. But all of it shares the same general climatic pattern that has existed for a long period of time geologically, producing a vegetation quite characteristic and often spoken of as a Mediterranean type. The moisture comes overwhelmingly in the cooler winter months and is followed by a long, dry period that is very hot toward the interior and cooler only near the coast, where the fogs and humidity of the ocean air help to prolong the growth season much more than in the hot interior. In either instance, at lower altitudes, snow falls in small amounts or not at all in winter, the flowering season is in spring, and there is little or no bloom in summer except along streams or about seeps and ditches. In the yellow pine belt and above, there is winter snow and the seasons are more like those in our more eastern and northern states.

This book deals largely with the area below the yellow pine and extending westward to the coast. It is an area of variable precipitation, from about 10 inches in the neighborhood of San Diego and parts of the Central Valley to about 100 inches in the extreme northern Coast Ranges. Usually, grassland prevails where the rainfall is from six to 20 inches; shrubby growth, chaparral, or scrub prevail in areas of rainfall from 15 to 25 inches; woodland is found where the rainfall is from 20 to 40 inches; and denser forest occurs in areas with higher rainfall, especially nearer the coast where the air is cool. These plant formations are not sharply separated by precipitation but often are by topography. Gently rolling hills may have grassland and, with a little more moisture, open woodland, whereas chaparral or other brush may appear on nearby stonier and steeper slopes.

Our broad-leaved evergreen trees and shrubs such as oaks (*Quercus* spp.) and California-lilacs (*Ceanothus* spp.) tend to have very harsh leaves with rather reduced surfaces as compared with their relatives in regions with summer rains, thus cutting down evaporation. Others may lose their leaves in the dry season, as does the California buckeye *(Aesculus californica)*. Still others, like the big-leaf maple *(Acer macrophyllum)*, grow only where their roots have access to moisture at all seasons. Overall, our California conditions produce much open country that becomes green with the advent of the rains in late fall or early winter. Seedlings of flowering annuals develop slowly through winter as does the new growth on shrubs and trees. The greatest season of flowering is from February to April or even May. Then brownness and dormancy again set in, and summer is largely a period of inactivity.

How to Identify a Wildflower

For identification, it is most helpful to have flowers available and not just the vegetative parts of the plant. To refresh your memory, the parts of a typical flower are shown in the illustration and briefly defined here. We begin with the outer usually greenish sepals, known collectively as the calyx. The inner usually colored petals together constitute the corolla. Next, we find the stamens, each typically with an elongate basal portion, the filament, and a terminal more saclike part in which the pollen is produced, the anther. In the center, the pistil has a basal enlarged ovary containing the immature seeds. Above the ovary is an elongate style, and one or more terminal stigmas, on which the pollen grains fall or rub off on an insect or hummingbird. These many parts may be greatly modified. The sepals may be separate, more or less united, and alike or not alike. The same is true of the petals. The corolla may consist of separate, similar petals. Petals may be reduced or quite lacking, or they may be united to form tubular, often two-lipped, structures that afford landing platforms

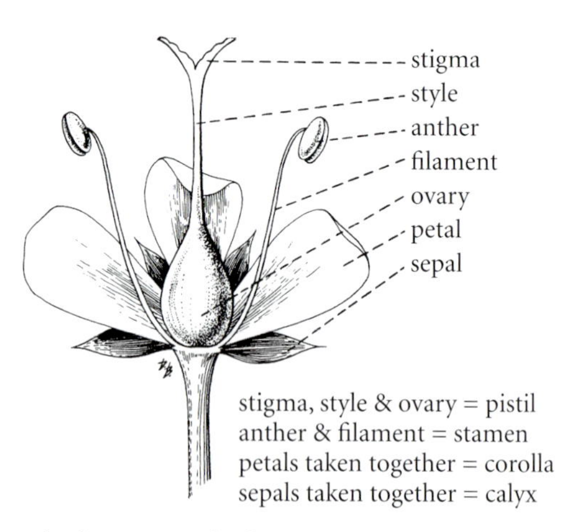

The parts of a representative flower

for bees and other visitors, in which case the stamens and style may be arched over so as easily to deposit pollen on or receive it from the body of the insect. The ovary may be partly sunken into tissues below or fused with them in such fashion as to be evident below the flower instead of up in it. When you look at a flower, you should observe such conformations and should pay some attention to the number of parts of a given series, as the petals. Superficially, the blue flower of a gilia (*Gilia* spp.) may resemble that of a *Brodiaea,* but the former shows five petal parts, and the latter has six segments. In other words, it is necessary at times to examine flowers in detail and with care.

The wildflowers illustrated by line drawings or color photographs in the text are arranged in four sections: (1) white to pale cream or pale pink or greenish white; (2) rose to purplish red or brown; (3) blue to violet; and (4) greenish yellow or yellow to orange. The fifth section describes various trees with attractive spring blooms. Identification of a red flower, for example, can be attempted by leafing through the illustrations in "Reddish Flowers" and by reading the text accompanying the illustrations. Such an artificial arrangement is helpful but does not always work. A lupine flower that is bluish when young and fresh often changes to reddish as it grows old and is about to die. Then, too, individuals of the same species can vary greatly in color, say, from blue to lavender or almost white. Almost every species has albino forms—white variants for red and blue types, yellow for those normally red or scarlet. Remember that you must allow for variation in color, for change with age, and for the fact that in nature growth is not fixed but fairly inconstant.

Perhaps a word should be added about the richness of the California flora. The long, dry season when dormancy is the rule is accompanied in the wet part of the year by a great wealth of annuals, many of which are highly colored. Since David Douglas came to the state more than a century ago to seek bulbs and seeds for introduction into England, our plants have been much prized in northern Europe. They have done well there because of the cool nights like our own and have been much used. Clarkias (*Clarkia* spp.), baby-blue-eyes *(Nemophila menziesii)*, and gilias, for instance, commonly met in gardens there, have often been greatly developed horticulturally, with many color forms, double flowers, and the like. But with the increase in population in California and with the encroachment on wild lands by industry and agriculture, many California wildflowers are increasingly rare. They need protection if we and our children are to enjoy them. Turning the camera on them rather than picking them is the means to permanent enjoyment.

Many of the plants that now seem to us a natural constituent of the California landscape were not here when the first Europeans arrived. I refer to the mustards (*Brassica* spp.), filarees (*Erodium* spp.), wild oats (*Avena* spp.), and many others, some of which may be quite weedy and unattractive (tumbleweed *[Salsola tragus]*, purslane *[Portulaca oleracea]*) or may be showy and add color to our fields and orchards (mustard, Bermuda-buttercup *[Oxalis pes-caprae]*, foxglove *[Digitalis purpurea]*).

Most of the drawings used in this book are by Dr. Stephen S. Tillett of the New York Botanical Garden; a smaller number are by Professor Richard J. Shaw of Utah State University.

To the foregoing people and to Gladys Boggess, secretary at the Botanic Garden, I express with great pleasure my gratitude for their help in securing materials and in preparation of the manuscript.

Philip A. Munz
Rancho Santa Ana Botanic Garden
Claremont, California
June 1961

INTRODUCTION TO THE PLANT COMMUNITIES OF CALIFORNIA IN SPRING

Robert Ornduff

California has long been known for its spring wildflower displays. In 1868, naturalist John Muir described the view that he witnessed as he looked eastward from the summit of Pacheco Pass, southeast of San Francisco Bay. He wrote, "at my feet lay the Great Central Valley of California, level and flowery, like a lake of pure sunshine...one rich furred garden of yellow compositae. And from the eastern boundary of this vast golden flower-bed rose the mighty Sierra, so gloriously colored and so radiant...". When Muir wrote those words, the human population of California was perhaps counted in thousands, whereas today it is counted in the tens of millions. Most of that "vast golden flower-bed" has been replaced with orchards, cultivated fields, freeways, and burgeoning towns and cities. But even today, one can find valleys, foothills, and mountain areas where the plant species that were seen by Muir well over a century ago still flourish.

The region covered by this book has a Mediterranean climate, with relatively low annual rainfall, moderate year-round temperatures, and prolonged summer drought. This climate is familiar to Californians because most of us live in an area with such a climate. Summer visitors from elsewhere find it unusual that our summers are so dry and that our hills turn brown in early summer. But if they visit in spring, they see vivid green hills and showy displays of colorful wildflowers that they cannot see anywhere else in the United States. This book enables you to learn the names of many of the spring wildflowers that are treasured by so many residents and visitors to California.

A recent estimate recognizes just over 4,800 species of native vascular plants in California. A vascular plant has specialized internal cells that conduct water, dissolved minerals and foods, and other substances throughout its internal tissues. Ferns, club-mosses, wildflowers, and trees are vascular plants; mosses and liverworts are not vascular plants and are not covered by this book, nor are ferns and club-mosses. An additional 1,000 species of introduced vascular plants occur in the state. Some introduced wildflowers are now very common in California, and a few of these are included in this book. A few of these introductions, such as field mustard *(Brassica rapa)*, filarees (*Erodium* spp.), and Bermuda-buttercup *(Oxalis pes-caprae)* (a garden pest detested by home gardeners) add welcome splashes of color to the late winter Californian landscape. They are here to stay indefinitely, whether we like them or not.

About three-quarters of California's native vascular plant species are perennials, which grow and flower over a period of several years. The rest of our native species are annuals, plants that complete their life cycle in a year or less. This high proportion of annuals in a flora (over 25 percent) is much higher than in any of the other four regions of the world with a Mediterranean climate (i.e., central Chile, southern Australia, the western cape region of South Africa, and lands around the Mediterranean Sea). Most of the species in Muir's "golden flower-bed" in the Central Valley are annuals, and the showiest spring displays in California are chiefly, but not exclusively, produced by annual species.

Most of our native annuals are winter annuals whose seeds germinate after the first heavy fall rains. The young plants grow until the onset of longer and warmer spring days, when they come into flower. As the soil dries after the rains have stopped, they produce seeds and then die and are represented only as seeds in or on the soil during the dry summer. Plants with this behavior are called drought evaders because they have no special morphological or physiological adaptations that allow them to survive prolonged drought. Instead, they persist through the dry season as dormant seeds.

Spring wildflowers can be found in many habitats and plant communities. California's best-known spring wildflower displays are in the deserts (not covered by this book) and in the grasslands of the Central Valley and the surrounding foothills that are included in this book. Unfortunately, the valley grasslands mostly have been converted to agriculture or have been urbanized, and the plant cover of only a few areas is reasonably intact. These are mostly in reserves and include such well-known wildflower localities as Table Mountain in Butte County, Bear Valley in Colusa County (still in private hands), and the Carrizo Plain in San Luis Obispo County.

A distinctive habitat that occurs in the Central Valley and in smaller adjacent valleys is called a vernal pool. Vernal pools occupy shallow depressions that are underlain by an impervious hardpan or clay. After the fall rains have begun, these depressions gradually fill with water, sometimes filling by mid-November or earlier. Vernal pools range in area from only a few square feet to those occupying many acres. During late autumn and winter, seeds of annual species that grow in these pools germinate and produce plants that come into flower once spring arrives. As the water level in the pools drops after the rains have ceased, different species come into flower sequentially, forming colorful concentric bathtub rings around the pool margins. First to flower include two members of the sunflower family: yellow goldfields (*Lasthenia* spp.) and bicolored yellow-and-white tidy-tips *(Layia platyglossa)*. These two wildflowers are not strictly vernal pool plants but often grow on the drier soil above the pools' high water level. Next to flower are various species of white or white and yellow meadowfoam (*Limnanthes* spp.). Once standing water has disappeared, downingias (*Downingia* spp.) produce their bright blue flowers on the pool bottoms. Because vernal pools are aquatic habitats during winter and completely dry during summer, they have not been invaded by introduced plants to the extent that is true of the surrounding grasslands. In these grasslands, introduced plant species often outnumber native ones. California's vernal pools also have a high proportion of species that occur only in the state but not elsewhere. Perhaps as many as 90 percent of our vernal pools have disappeared, mostly because of agricultural activities. Areas where they can still be seen include Vina Plains in Tehama County, Santa Rosa Plain in Sonoma County, Jepson Prairie in Solano County, Big Table Mountain in Fresno County, and the San Diego area. Adjacent pools often support somewhat different species, so be sure to visit as many pools as you can while in the area.

Boggs Lake, in Lake County, is an unusual vernal pool because it is located at about 2,000 feet and occurs in a woodland rather than in grassland setting. Portions of the lake that retain water or moist soil through summer support an array of perennial marsh plants. The lake margin that becomes dry during summer hosts an unusual assortment of annual wildflowers, some of which are rare and a few of which occur only here. Because of its elevation, the wildflower displays at Boggs Lake are in late spring and early summer, a time when the valley pools are dry. Vernal pools (or vernal lakes) that have various levels of salinity and characteristic floras are found at the Grassland Ecological Preserve in Merced and the vast Carrizo Plain in San Luis Obispo County, where Soda Lake becomes an enormous vernal pool during wet winters.

Accounts of early explorers in California and other historical records indicate that well into the nineteenth century, much of the Central Valley was marsh or occupied by standing water during most or all of the year. Indeed, at one time the largest freshwater lake in the United States west of the Mississippi River was in the San Joaquin Valley. It was called Tulare Lake and supported a thriving fishing industry. It disappeared in the nineteenth century as its tributary water was diverted to agricultural uses. Today, the Central Valley still supports large areas of freshwater or brackish marshes. Some of these are remnants of marshes that predate the development of agriculture in the valley, and others occur where water is impounded or channeled as a result of agricultural activities. Because freshwater marshes contain standing water, moist soil, or both, during the entire year, they support a group of plant species different from those found in vernal pools. Most freshwater marsh plants are perennial and herbaceous, and many have good powers of vegetative reproduction. Examples of wildflowers that grow in freshwater marshes are fringed water-plantain *(Damasonium californicum)*, broad-fruited bur-reed *(Sparganium eurycarpum)*, cow-lily *(Nuphar luteum* subsp. *polysepalum)*, and, found along the central and northern coast, yellow skunk-cabbage *(Lysichiton americanum)*. Despite its uncomplimentary name, this last species is elegant in flower and in leaf. I consider its floral odor (in low concentrations) to be pleasant. Many common freshwater marsh species have inconspicuous flowers and are not included in this book.

Several attractive or otherwise interesting wetland wildflowers in the region covered by this book do not grow in Central Valley marshes but grow in damp meadows or along year-round streams or rivers outside the valley. These include umbrella plant *(Darmera peltata)*, western azalea *(Rhododendron occidentale)*, California lady's-slipper *(Cypripedium californicum)*, camas *(Camassia quamash)*, California pitcher plant *(Darlingtonia californica)*, various monkeyflowers (*Mimulus* spp.), stream orchid *(Epipactis gigantea)*, and golden-eyed-grass *(Sisyrinchium californicum)*.

Many spring wildflowers occur only in specialized habitats. Vernal pools have already been mentioned, but soil type is also an important influence on plant distribution. In various areas in the Sierra foothills and the Coast Ranges there are small to extensive outcrops of serpentine rock. Freshly exposed serpentine typically is brittle, grayish green, smooth, mottled, and shiny. When weathered, especially in wetter regions, it may turn reddish. Serpentine is iron magnesium silicate and has toxic levels of chromium and nickel and low levels of calcium, nitrogen, molybdenum, and other elements that are essential for normal plant growth. Soils derived from serpentine consequently have a poor nutrient status, as well as toxic qualities. One might guess that few plants can survive on serpentine, but in fact serpentine supports a number of interesting plant species, many of which grow on it only. Serpentine rock and the soils derived from it generally support a vegetation type different from that on adjacent soils of better quality. For example, serpentine outcrops may support chaparral or grassland in areas otherwise occupied by woodlands or forests. In spring, serpentine areas are worth a visit. Bitter root *(Lewisia rediviva)* is often found on serpentine in regions where it does not otherwise occur. Jewelflower *(Streptanthus glandulosus)* and its close relatives are common on serpentine. Streams that run through serpentine are often lined with thickets of the fragrant western azalea, and lucky individuals may be able to find occasional clumps of California lady's-slipper or California pitcher plant.

The hills on both sides of the Central Valley support patches of chaparral, a vegetation type dominated by shrubs. Chaparral commonly grows on poor soils (including serpentine). Chaparral shrubs typically are dense and much branched and have hard-textured evergreen leaves that often are covered with wax to reduce water loss. Chaparral is highly fire prone, and in many areas it requires periodic fires to maintain itself. Whereas some chaparral shrub species are killed by fires and must reproduce via seeds, others have burls or large woody tubers just below the ground surface and produce new shoots from these soon after a fire has swept through. The first spring following a chaparral fire, wildflower displays are often impressive, and a few annual wildflower species are known to occur only after fire. Manzanitas (*Arctostaphylos* spp.) and chamise *(Adenostoma fasciculatum)* are common shrubs in chaparral. Also present are various species of silk tassel (*Garrya* spp.), yerba santa *(Eriodictyon californicum)*, species of ceanothus (*Ceanothus* spp.), and bush poppy *(Dendromecon rigida)*.

Chaparral often occurs as ecological islands in the oak woodlands of the foothills around the Central Valley. These woodlands mostly occur between the yellow pine forest and the valley floor. The predominant trees are widely scattered live (evergreen) oaks and deciduous oaks with occasional shrubs and extensive grassy areas among the trees. California botanist W.L. Jepson wrote that the oaks in these woodlands play "a strong and natural part in the scenery of the yellow-brown foothills. Always scattered about singly, or in open groves, the trees are well associated in memory with bleached grass, glaring sunlight, and dusty trails, although for a few brief days at the end of the rainy season the…trunks rise everywhere from a many-colored cloth woven from the slender threads of innumerable millions of flowering annuals." These annuals include cream cups *(Platystemon californicus)*, lupines (*Lupinus* spp.), popcornflower (*Plagiobothrys* spp.), buttercups (*Ranunculus* spp.), Chinese houses *(Collinsia heterophylla)*, bird's eyes *(Gilia tricolor)*, fiddlenecks (*Amsinckia* spp.), and many colorful perennials.

The Coast Ranges and immediate coast support some vegetation types that are not represented in the Sierra Nevada. Moving coastward from the foothill region occupied by oak woodlands, one encounters a mixed evergreen forest, a forest that occurs in hilly areas on the drier margins of the dense coniferous forests found to the west along the wetter coast. The term "evergreen" refers not to conifers, but to hardwood trees that bear leaves year-round. These forests include Pacific madrone *(Arbutus menziesii)* and several trees not described in this book: California bay *(Umbellularia californica)*, bigleaf maple *(Acer macrophyllum)*, tanbark oak *(Lithocarpus densiflorus)*, and several evergreen and deciduous oak species.

The north coastal forest occurs in areas along the coast with rich, deep soils, fairly high rainfall, and summer fog. Its dominant trees are not described in this book but include the familiar coast redwood *(Sequoia sempervirens)* and a number of other conifer species that form dense, dark forests. Few annual wildflowers grow in these forests, but there are many colorful shrubs and perennial herbaceous species to be found there. These include inside-out flower *(Vancouveria planipetala),* milk maids *(Cardamine californica),* woodland star (*Lithophragma* spp.), salal *(Gaultheria shallon),* Pacific starflower *(Trientalis latifolia),* yerba buena *(Satureja douglasii),* coltsfoot *(Petasites frigidus* var. *palmatus),* long-tailed wild-ginger *(Asarum caudatum),* redwood-sorrel *(Oxalis oregana),* fairy bells (*Disporum* spp.), and trilliums (*Trillium* spp.). A particular favorite of mine, an indicator that spring will soon arrive, is a plant Munz called brownies *(Scoliopus bigelovii),* which has small, intricate brownish and greenish flowers. This common plant of deep shade often flowers soon after Christmas. Although Munz, a resident of southern California, used the name brownies for this plant, Jepson says this name is limited to Humboldt County. Instead, Jepson calls it slink pod (as I do), an allusion to the fact that as the seed capsules form, their stalks elongate and push them away from the mother plant. The species is also called fetid adder's tongue, a reference to the apparently bad odor of its strange flowers that are pollinated by flies. I confess that I have never gotten my nose close enough to these flowers to appreciate their perfume.

To the coastward side of the north coastal forest on slopes and flatlands that are exposed to strong sea winds is coastal scrub, dominated by a number of shrub species, most of which are not described in this book except for salal but are extensively covered in Munz's *Shore Wildflowers.* On the central and northern California coast, these scrublands form a mosaic with coastal prairie, dominated by grasses and other herbaceous species. The shifting coastal dunes are occupied by coastal beach and dune plants, and occasional coastal salt marshes support their own array of plant species, most of which do not have showy flowers. Distinctive plants found in coastal salt marshes include glasswort *(Salicornia subterminalis),* western sea-lavender *(Limonium californicum),* and gumplant *(Grindelia stricta).*

In open grassy or sandy areas along the coast, the wildflower displays may rival those of the Central Valley. At Point Reyes, for example, there are large patches of goldfields (*Lasthenia* spp.), various lupines, fiddlenecks, iris *(Iris douglasiana)*, man-root (*Marah* spp.) (interesting in fruits as well as in flower), sand-verbenas (*Abronia* spp.), buttercups, beach evening-primrose *(Camissonia cheiranthifolia)*, the introduced sea rocket *(Cakile maritima)*, sea-fig *(Carpobrotus chilensis)*, checker mallow *(Sidalcea malvaeflora)*, thrift *(Armeria maritima* subsp. *californica)*, California figwort *(Scrophularia californica)*, seaside daisy *(Erigeron glaucus)*, California poppy *(Eschscholzia californica)*. Next to the road to Drake's Beach is a large population of California poppy that is of hybrid origin between the distinctive low, gray-leaved coastal form with bicolored flowers and the more inland tall, green-leaved, orange-flowered form apparently brought into cultivation by a nearby rancher. Golden-eyed-grass *(Sisyrinchium californicum)*, Pacific silverweed *(Potentilla anserina* subsp. *pacifica)*, and sun cup *(Camissonia ovata)* are also found at Point Reyes. The flowering season along the coast extends throughout the year. As I write this during Thanksgiving week, field mustard is beginning to transform some of the fields at Point Reyes into sheets of butter yellow. During autumn, one can always find a few flowers on many coastal plants that flowered profusely much earlier in the year. Does November qualify as spring? For the annual field mustard that is bursting into flower it is spring.

The California spring wildflowers that we find growing together are mixtures of species that have had diverse historical origins. Millions of years ago, California had higher rainfall and a cooler climate than it does today. The wetter, cooler northern portion of the state supported forests that contained a large number of coniferous tree species belonging to several genera, as well as numerous hardwood tree species. The climate has changed over the past 15 to 20 million years in western North America, with an overall trend toward warmer annual temperatures, lower rainfall, and loss of summer rainfall. Ultimately, conditions became too unfavorable to support the growth of many tree species that grew here. Many of these, especially hardwoods, disappeared entirely from western North America. Nearly all of them, including magnolia *(Magnolia)*, elm *(Ulmus)*, sweet gum *(Liquidamber)*, ginkgo *(Ginkgo)*, hickory *(Carya)*, bald cypress *(Taxodium)*, beech *(Fagus)*, persimmon *(Diospyros)*, dawn redwood *(Metasequoia)*, and chestnut *(Castanea)* still persist in eastern North America, Eurasia, or both. Others, such as coast redwood, became restricted to wetter sites, such as north-facing slopes and valleys and are still here today. In California, conifers are now abundant in moister coastal and montane sites, and the number of hardwood tree species found here is much smaller than formerly.

About half the native plant species found in California today are descended from members of a northern flora that once extended more or less continuously around the temperate northern hemisphere. Most of our conifers and some hardwood trees such as mountain dogwood *(Cornus nuttallii)* and California buckeye *(Aesculus californica)* belong to this flora. Many herbaceous plants are members as well, including fawn-lilies (*Erythronium* spp.), California saxifrage *(Saxifraga californica)*, columbines (*Aquilegia* spp.), peonies (*Paeonia* spp.), Indian-pink *(Silene californica)*, wild-ginger (*Asarum* spp.), redwood-sorrel *(Oxalis oregana)*, Indian warrior *(Pedicularis densiflora)*, irises, larkspurs (*Delphinium* spp.), and trilliums. Close relatives of all these trees and wildflowers still occur in eastern North America, Europe, and Asia.

During the Miocene epoch, beginning about 25 million years ago, much of central and southern California was covered by a dry woodland containing live oaks (*Quercus* spp.) growing with palms (Arecaceae), hollies (*Ilex* spp.), laurels (Lauraceae), and so forth, and an understory of shrubs that were the ancestors of the modern chaparral shrubs. This dry-adapted vegetation type is believed to have migrated into California from the mountains of northwestern Mexico. As conditions in California became drier, the trees disappeared from these woodlands, leaving the understory shrubs that eventually developed into our familiar chaparral. About one-third of the native species of California are considered to be descendants from this immigrant flora. These include manzanitas, ceanothus, and western sycamore *(Platanus racemosa)*. Other California plants have close relatives in the Mediterranean region, such as snowdrop bush *(Styrax officinalis* var. *redivivus)*, snapdragon (*Antirrhinum* spp.), and western redbud *(Cercis occidentalis)*. Still others are of tropical origin. Many are descended from inadvertent introductions that have arrived in California since the Spanish Mission period that began in 1769. A few of these introductions have been mentioned above and produce colorful spring floral displays. Some of these introductions do not appear to have displaced native plants (such as foxglove *[Digitalis purpurea]*) but others, such as tamarisk (*Tamarix* spp.), have become serious pests along watercourses and have crowded out native species in these habitats.

A few of our wildflowers have unusual geographic distributions. Among these are beach strawberry *(Fragaria chiloensis)*, which grows naturally along the Pacific Coast of North America, at about 7,000 feet on the Big Island of Hawaii, and on the coast and in the Andes of Chile and Argentina. Although the most common of our coastal sea-figs, or iceplant *(Carpobrotus edulis)* is an introduction from South Africa, our other sea-fig *(Carpobrotus chilensis)* is a native that also occurs along the coast of central Chile. Sea-figs otherwise are native to the Cape region of South Africa and to Australia. Our spring wildflowers can be enjoyed not only for their beauty, but also for the interesting stories that each has to tell.

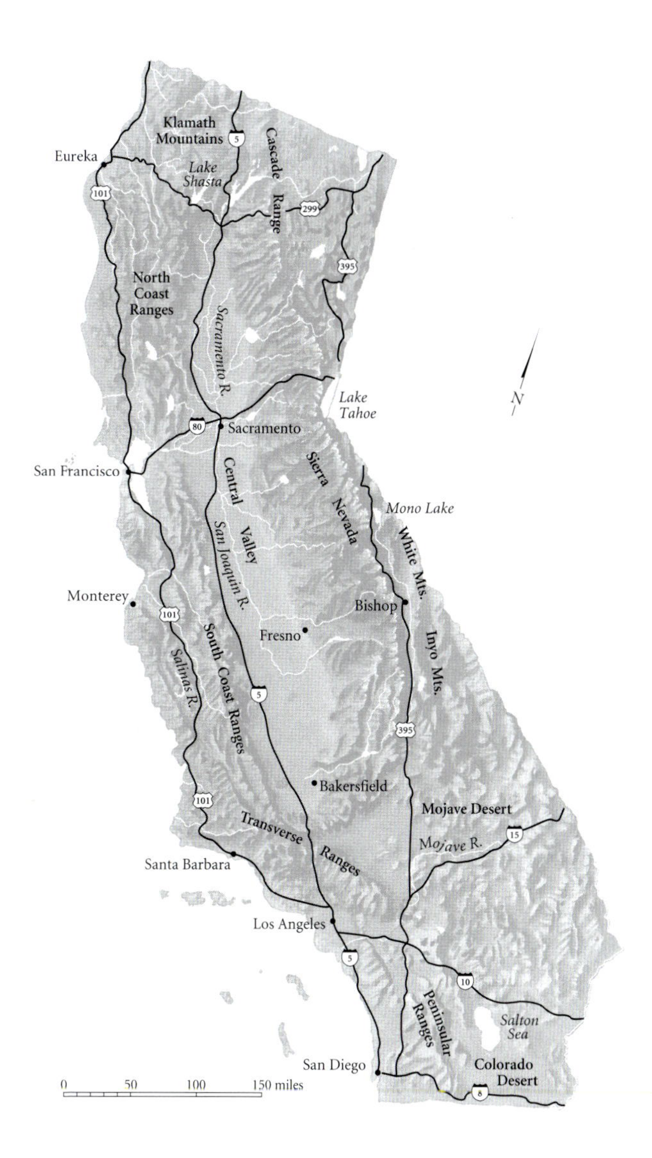

Maps of California

WHITISH FLOWERS

White to Pale Cream or Pale Pink or Greenish White

FRINGED WATER-PLANTAIN, or **STAR WATER-PLANTAIN**, ***(Damasonium californicum)*** is an attractive wetland plant with a woody base, a cluster of long-stemmed, narrow leaves, and tall flower stalks that bear several white, three-petaled flowers with toothed to deeply cut petals. Each flower has six stamens and six to 15 pistils in a whorl, and the six-parted, long-beaked fruits form star-shaped clusters. A member of the water-plantain family (Alismataceae), this plant has its roots in the water, but the upper parts are emergent. Another common species of this family is common water-plantain *(Alisma plantago-aquatica)*, in which the flower petals are entire or only slightly cut, and the fruits are erect rather than spreading and have only very short beaks. Also common is broad-leaved arrowhead *(Sagittaria latifolia)*, often called tule-potato, or wapato, with arrow-shaped leaves and flower clusters bearing only male flowers with stamens in the upper parts and seed-producing female flowers in the lower parts. All three plants can be found in marshes and other wetland areas, with fringed water-plantain limited to central and northern California, and the other two species occurring throughout the state.

Fringed water-plantain, or star water-plantain

Broad-fruited bur-reed

BROAD-FRUITED BUR-REED ***(Sparganium eurycarpum),*** in the cattail family (Typhaceae), is another common plant with its roots in the water and the upper parts emergent. It differs from the above species, however, in having leaves up to 40 inches long and flowers with no petals. The upper rounded flower clusters bear six to 14 short-lived male flowers with one to eight stamens each, and the lower clusters have one to 20 fertile female flowers that produce brown, one-seeded, burlike fruits that look like round, spiked balls. This species of bur-reed can be found along the edges of lakes and ponds, as well as in marshes and along streams, throughout most of California.

A very large family in California is the lily family (Liliaceae), which contains a variety of different-looking plants. The flowers are always six parted, often with the sepals and petals resembling each other, and the leaves are usually long and thin, parallel veined, and often basal or whorled around the stem. Many of the plants in this family grow from bulbs or corms (underground stems), and **SOAP PLANT**, or **AMOLE**, ***(Chlorogalum pomeridianum)*** is a typical member of the family in this regard. It has a large underground bulb that is encased in numerous dark brown fibers left over from old bulb coats. A conspicuous basal cluster of long, thin, wavy, blue green leaves appears after the first fall rains, and in middle to late spring a flower stalk begins to grow that eventually reaches a height of five to eight feet. Numerous small white flowers are borne on the tall branched stems, and each flower is about an inch long or less and consists of three reflexed sepals and three reflexed petals. Pollinated by moths, the flowers do not open until dusk, and they close up before the next morning, so only the evening hiker sees this plant in bloom. Soap plant is found in dry, open places below 5,000 feet from San Diego to southern Oregon. Native Americans roasted and ate the starchy bulb, and early settlers commonly used it as a shampoo.

ZYGADENE-LILY, or **STAR-LILY**, ***(Zigadenus fremontii)*** is another bulbous plant in the lily family with narrow basal leaves and a

Zygadene-lily, or star-lily

branched flower stalk. The leaves, however, are flat and green and scattered up the stem as well as clustered at the base, unlike soap plant. The flower stalk is shorter, the flowers are larger and more crowded together on the stem, and the inflorescence is often compound below but simple above. The numerous small white to yellowish star-shaped flowers have shiny greenish yellow glands near the center, and the petals are wider and spreading, not reflexed as in soap plant. Found on grassy or bushy slopes and often more abundant after a fire, it occurs below 3,500 feet from San Diego County to Oregon. The bulb of zygadene-lily is poisonous, so it is important to distinguish it from the edible soap plant. In addition to the leaf and flower differences described, the bulb has no fibers surrounding it.

Slim Solomon's-seal

The genus false Solomon's-seal *(Smilacina)*, also in the lily family, occurs from the Pacific Coast to the Atlantic. In California, **SLIM SOLOMON'S-SEAL *(Smilacina stellata)*** is the most common species. It is a perennial herb with thick creeping underground rootstocks and ascending stems with two rows of lanceolate to ovate leaves and a terminal cluster of five to 15 small, white, star-shaped flowers followed by reddish or red purple berries. False Solomon's-seal *(S. racemosa)* is also common but much more robust, taller, and has larger leaves and a long dense cluster of many small flowers. Both species are found through most of California, especially in partly shaded places below 8,000 feet.

Related to false Solomon's-seal *(Smilacina)* is **HOOKER'S FAIRY BELLS *(Disporum hookeri)*** with slender rootstocks and leafy stems usually thinner and more crooked than the above. It has only one to three large hanging flowers instead of elongate clusters, and the narrowly bell-shaped flowers are greenish to white and about half-an-inch long and have long sta-

Hooker's fairy bells

mens extending beyond the flowers. Scarlet berries appear later in the year. This plant is found in shaded woods away from the immediate coast up to about 5,000 feet from Monterey County to southern Oregon. Also common is large-flowered fairy bells *(D. smithii)*, usually a taller, more robust plant with larger, white flowers and stamens shorter than the petals. It is also found in shaded woods but is most common in redwood forests.

SIERRA FAWN-LILY *(Erythronium multiscapoideum)* grows from an elongated bulblike structure. It has two mottled leaves near the base of the stem and a cluster of one to 10 elegant, nodding flowers up to an inch-and-a-half long. The white petals are reflexed and have yellow bases. Growing in open woodlands between 1,000 and 3,000 feet in the north and central Sierra Nevada and in the Cascade Ranges, it blooms from March to May. Other species of fawn-lily in California have flowers ranging from white to yellow to rose, and several

Sierra fawn-lily

species have no mottling in the leaves. The size of the flower and the shape of the stigma also help distinguish the various species from one another.

One of the most beautiful groups of California wildflowers is the genus mariposa-lily *(Calochortus)*, also in the lily family. Two basic types occur: fairy lantern or globe-lily, with small, closed, pendent flowers; and mariposa-lily, with larger, erect, bowl-shaped flowers. **WHITE FAIRY LANTERN**, or **WHITE GLOBE-LILY**, ***(Calochortus albus)*** in the first group, is one of the most delicate of wildflowers with its stem rising from a basal group of two to six long thin leaves and bearing a nodding cluster of rounded white

White fairy lantern, or white globe-lily

flowers, each about an inch long. A single long thin leaf precedes the flowering stalk earlier in spring, letting the hiker know where this lovely flower will appear in about a month's time. It grows in shaded, often rocky, places in woods and canyons below 7,000 feet from San Diego County to the San Francisco Bay Area and in the foothills of the Sierra Nevada. In the Santa Lucia Mountains and other places in southern California it takes on a rose tinge and can look quite different.

Butterfly Mariposa-lily

Mariposa-lily is the second and larger group in the mariposa-lily *(Calochortus)* genus with erect, open, more or less bowl-shaped flowers. **BUTTERFLY MARIPOSA *(Calochortus venustus)*** has underground bulbs, long linear leaves, and large, generally white flowers with a multicolored blotch in the center of each of the three petals and usually a second paler blotch above. The shape and texture of the nectary at the base of each petal helps distinguish the different species of mariposa-lily, being more or less square in this species. It grows above 3,000 feet in the Sierra Nevada and in the Coast Ranges from the San Francisco Bay Area northward. It begins to bloom in May and is replaced at lower elevations from the San Francisco Bay Area southward by a very similar species that is often mistaken for it, clay mariposa-lily *(C. argillosus),* which often lacks the second spot on the petals and has a crescent-shaped nectary. The two species can look remarkably similar, and there can be much variation in both color and design. Thus, elevation is perhaps the easiest way to distinguish them, with butterfly mariposa growing above 3,000 feet and clay mariposa-lily growing below 3,000 feet.

A large group of plants within the lily family has flowers in umbels, that is, the flowers radiate out from one central point at the top of the stem rather than being scattered along it. This group has often been separated out and placed in its own family, the onion family (Amaryllidaceae), but currently it is included within the lily family. Daffodils (*Narcissus* spp.) and lily-of-the-Nile *(Agapanthus orientalis)* are cultivated examples of this group, and onions *(Allium)* are also in this group. Another native member, and strongly resembling onion, is muilla ("allium" spelled backward), but it lacks the onion taste and odor. The most widespread species is **COMMON MUILLA *(Muilla maritima).*** It has greenish to white flowers with identical petals and sepals, each with a brown stripe down the center. The stamens have blue, green, or purple anthers. It is found up to 7,500 feet on flats and slopes in the Coast Ranges and occasionally in the Sacramento and San Joaquin Valleys, the cen-

Common muilla

tral Sierra Nevada, and the western desert areas. A very similar plant is white brodiaea *(Triteleia hyacinthina)*, which is usually somewhat taller and has a green stripe down the center of each petal and sepal, rather than brown.

YUCCA, or **OUR LORD'S CANDLE**, ***(Yucca whipplei)*** is unmistakable, with its flowering stalk up to eight feet tall bearing innumerable white to purplish flowers an inch or more long and arising from dense clumps of stiff, swordlike leaves that can be up to four feet long. It occurs up to 8,000 feet in desert, chaparral, and coastal scrub communities in open places and in fans below canyons away from the immediate coast in the South Coast Ranges, southern Sierra Nevada, and southwestern California from Monterey and Tulare Counties southward. It used to dominate considerable areas, but because it is such an impressive plant, every car returning to the city after a Sunday excursion in the early 1900s had an eight-foot yucca flower stalk tied alongside, and it has thus become much less common. It is now protected by law, as are most native plants, and cannot be collected without permission. It flowers from April to June.

Yucca, or Our Lord's candle

WHITE TRILLIUM, or **WESTERN TRILLIUM**, ***(Trillium ovatum)*** is one of the few plants in the lily family that has broad, network-veined leaves instead of the more typical narrow, parallel-

White trillium, or western trillium

veined ones of this family. The three ovate leaves are in a whorl at the top of a stem that is usually four to 10 inches high but can sometimes be more than two feet tall. A thin delicate stem rises from the center of the whorl of leaves, and a single white or pale pink three-petaled flower sits at the tip. About one to three inches long, the flower may turn rose in age after it has been pollinated and starts producing fruit. Found on moist, shaded slopes in redwood and mixed evergreen forests, it occurs up to 6,500 feet from Monterey County northward, flowering as early as February.

In the orchid family (Orchidaceae), the flowers are also six parted with three petals and three sepals, but they are asymmetrical and often have a spur. The sepals may or may not resemble some of the petals, and the two upper petals are usu-

California lady's-slipper

ally quite different from the lower one, which is generally in the form of a lip or a pouch. **CALIFORNIA LADY'S-SLIPPER *(Cypripedium californicum)*** can be found dotting the forest floor with its yellowish to white flowers composed of two erect upper sepals and a less-conspicuous lower one, two spreading side petals, and a large white inflated lower petal. It is one to two feet high, and the three to 10 flowers are each almost an inch long. The five to 12 leaves are alternate on the stem, the lower ones being more or less oblong and the upper ones more tapered and lanceolate. It is found on wet, rocky ledges and hillsides below 5,000 feet from Marin County to southwestern Oregon and begins to bloom in May. Other species of lady's-slipper extend as far south as Santa Cruz and Mariposa Counties and may have a more greenish or purplish lip.

Yerba mansa, or lizard's tail

The only member in California of the lizard's tail family (Saururaceae) is **YERBA MANSA** ***(Anemopsis californica),*** meaning "soft herb," and it is often called by the family name, **LIZARD'S TAIL.** It is a peculiar plant with large, mostly basal, long-stemmed leaves and long runners (long flat stems along the ground) that produce new individuals. The flower stalks are about four to 20 inches tall and end in what looks like a flower with a number of large white petals near the base. Close examination, however, reveals that these are actually modified leaves called bracts. They become progressively smaller as they continue up the spike, and each of the smaller bracts sits below a petal-less flower with six stamens and an inconspicuous pistil. Found in low, wet, usually alkaline places up to 5,500 feet, its range extends from the Sacramento Valley and San Francisco Bay Area to Baja California and Texas.

Western wood anemone, or western windflower

The buttercup, or crowsfoot, family (Ranunculaceae) often has plants with deeply divided to compound leaves that resemble a bird's foot. The flowers generally have five petals (or sometimes none) and multiple stamens and pistils. **WESTERN WOOD ANEMONE**, or **WESTERN WINDFLOWER**, ***(Anemone oregana)*** is a small delicate plant in this family from four to 10 inches tall, with one or two whorls of deeply lobed or compound leaves. The single flower is less than an inch long and has no petals, but the five fragile sepals are petal-like and can be white, red, blue, or purplish. Occurring below 5,000 feet, it is a plant of coniferous forest slopes, especially redwoods, from Santa Cruz County northward to Washington, and is also found in the northern Sierra Nevada.

Another member of this family is **PIPESTEMS**, or **CLEMATIS**, ***(Clematis lasiantha),*** a woody climber 10 to 15 feet high that spreads out over bushes in chaparral and open woodlands. The compound leaves usually have three leaflets, and the large, hairy flowers are petal-less but have four large, spread-

Pipestems, or clematis

ing, very petal-like sepals, forming attractive masses of cream color. The numerous pistils produce densely hairy fruits in summer and fall that look like giant cotton balls clambering over the brush. Growing below 6,500 feet from central California and the Sierra Nevada southward to the Mexican border, it flowers between January and June. Very similar, but blooming and fruiting later and having five or more leaflets, is virgin's bower, or yerba de chiva, *(C. ligusticifolia)*, found in similar, but moister, habitats, usually near streams and other wet areas.

One of the loveliest of our spring flowers is **CREAM CUPS *(Platystemon californicus)*,** in the poppy family (Papaveraceae), a family that generally has radial flowers with four to six petals and two

Cream cups

to three sepals. This species is an annual with several long, thin opposite leaves on a very hairy stem. A single flower less than an inch long sits atop each four- to eight-inch-long flower stem. The six spreading petals, with three hairy sepals below, can be almost white to cream or yellow, occasionally with a reddish tinge. It is found in open sandy or clay areas on burns and disturbed places below 3,000 feet in most of California.

Coulter's matilija poppy

COULTER'S MATILIJA POPPY ***(Romneya coulteri)*** is one of our most elegant plants and has the largest flower of any California native. Growing treelike on rather woody stems three to seven feet tall, the several blossoms have three smooth sepals, six crinkled petals two to four inches long, and numerous yellow stamens. The leaves are grayish green and deeply lobed or divided into three to five main segments. It can be found blooming in May and June in dry washes and canyons below 4,000 feet in the Peninsular Ranges and the eastern parts of the South Coast Ranges. A similar species, hairy matilija poppy *(R. trichocalyx)*, has smaller flowers and leaves and sepals with long stiff hairs. It occurs in similar habitats in the western parts of the South Coast Ranges southward to Baja California. Both have become uncommon over the years because of various factors, including human impact and incursions of nonnative plants.

With white flowers like those of Coulter's Matilija poppy *(Romneya coulteri)*, but smaller and more crinkled, is **CHICALOTE**, or **PRICKLY POPPY**, ***(Argemone munita)***. Growing two to four feet high, it is less woody than the above species and is covered throughout with stiff spines. The leaf is up to six inches long and prickly on both surfaces. The plant exudes a yellow sap. It is found in various forms in open places below 10,000 feet in much of California and begins to bloom in

Chicalote, or prickly poppy

March. A related species, leafy prickly poppy *(A. corymbosa)*, occurs in the desert and on the eastern side of the Sierra Nevada. It exudes an orange sap, and the leaves are less prickly on the upper surface.

Redwood-ivy, or inside-out flower

REDWOOD-IVY, or **INSIDE-OUT FLOWER**, ***(Vancouveria planipetala)*** bears the name of Captain George Vancouver, who visited the California coast in 1792. In the barberry family (Berberidaceae), which is characterized by generally having its sepals and petals in two to three whorled sets, this delicate-looking plant is related to Oregon-grape *(Berberis aquifolium)* and California mahonia *(B. pinnata)*. A perennial herb with a running rootstock, it has ternately (three-parted) compound leaves on very thin, reddish brown stems. The small, white flowers are unique with their two sets of six sepals, one set erect and one reflexed, and a single set of six reflexed petals. It grows up to about 4,000 feet in the shade of woods from Monterey County northward.

Related to the poppies but with much simpler flowers is the mustard family (Brassicaceae), with a biting peppery sap and flowers with four petals and six stamens, two of the stamens usually appearing lower than the other four. Many members of this family, such as radish, cabbage, and cauliflower, are important food plants. Of the many California natives in this family, **HAIRY FRINGEPOD**, or **HAIRY LACEPOD**, ***(Thysanocarpus curvipes)*** is an unusual annual with slender erect stems and unique round, almost flat, seedpods with scalloped, often perforated, winged edges. The bases of the lanceolate leaves

Hairy fringepod, or hairy lacepod

clasp the stem, at least on the upper stem parts, and the tiny white to purple tinged flowers are inconspicuous and last only a short time. The distinctive, long-lasting round fruit pods are the distinguishing mark of this plant. It is common in grassy or brushy places below 6,000 feet from Baja California northward to British Columbia. Narrow-leaved fringepod *(T. laciniatus),* also common, looks similar but is more delicate and does not have clasping leaves. It is usually found in rockier areas.

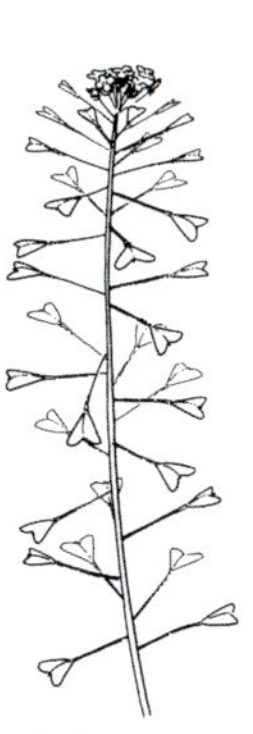

A small plant found in almost every field and vacant lot below 8,000 feet, and originally brought over from Europe, is **SHEPHERD'S PURSE *(Capsella bursa-pastoris),*** also in the mustard family. The tiny white flowers clustered at the top of the stem are inconspicuous, but as the stem elongates they spread out and produce distinctive heart-shaped fruits that can often catch the hiker's eye. When ripe, the fruit splits down the middle, revealing several orange yellow seeds—the "golden coins" in the shepherd's purse. In Europe, this plant is known as Mother's heart because of the shape of the fruit. It is usually only a few inches tall and has a basal rosette of leaves that can range from barely lobed to deeply dissected. The leaves along the stem are much smaller and have clasping bases. It is found in most disturbed or waste places throughout California. Also small, weedy, and inconspicuous is bittercress *(Cardamine oligosperma),* found in slightly moister places and often in gardens. It, also, has clusters of tiny white flowers at the stem tips, but as the stem elongates the flowers produce long, narrow, flat pods. The leaves are mostly in basal rosettes and are pinnately dissected.

The mustard family is divided into two basic groups distinguished by their fruit (seedpod). The above three plants are all in the round-pod group, whereas the following three plants, as well as mustard itself *(Brassica)*, are in the long-pod group. **YELLOW-FLOWERED THELYPODIUM** ***(Guillenia flavescens)*** is one of the more attractive members of this latter group. The common name is somewhat misleading because the showy, wavy-edged petals are actually white. The sepals, however, which are erect or spreading and pouched at the base, are creamy white to pale yellow, or even purple, and can give the flower a more yellowish appearance, especially in smaller-petaled plants. The basal leaves are often wavy-edged as well, and the upper leaves do not clasp the stem as they do in related species. The long narrow seedpods range from one to more than three inches long and can be ascending to reflexed. It occurs below 2,500 feet on dry, open slopes, often on serpentine, in the inner Coast Ranges the length of the state, in a few places east of San Francisco Bay, and in the Sacramento Valley.

MILK MAIDS, or **TOOTHWORT**, ***(Cardamine californica)*** is one of the earliest spring flowers, blooming as early as January or early February. The tuberlike rhizomes produce wide rounded rhizome leaves along the ground that are separate from the flowering stems, which have compound leaves divided into three to five ovate to oblong leaflets with wavy to dentate edges. The flowers have four milky white petals, each about half-an-inch long, and produce long, thin, erect seedpods. Usually found in shaded canyons and woodlands through most of California

Milk maids, or toothwort

up to 4,000 feet, there are five different varieties distinguished mainly by the shape and composition of the rhizome leaves.

The purslane family (Portulacaceae) is usually easily recognized by its flowers with five or more petals but only two sepals. Some members of the *Lewisia* genus, however, can have as many as eight sepals, such as **BITTER ROOT *(Lewisia rediviva),*** which has six to eight petal-like sepals and 10 to 19 long, thin petals. It has a thick taproot and a rosette of fleshy leaves that almost blend into the gravel or talus where this plant grows. The large, showy white flowers are in tufts on the ground and often turn pink as they age. Growing up to 10,000 feet from the San Jacinto Mountains northward, it is also found in other parts of the northwest and is the state flower of Montana. (See photograph, page 44.)

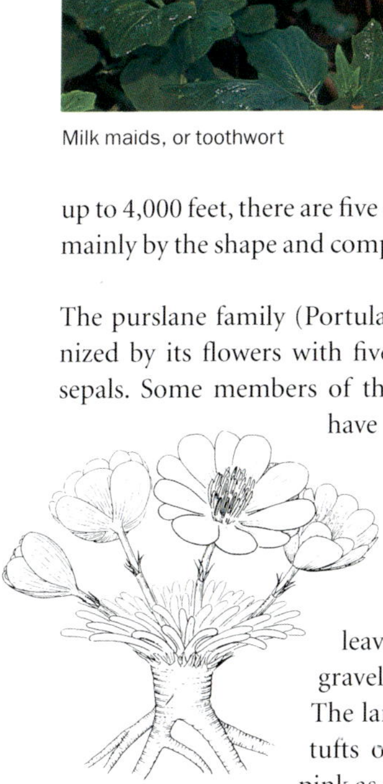

Bitter root

MINER'S-LETTUCE ***(Claytonia perfoliata)*** is probably the most common and widespread plant in the purslane family. It is easily recognized by its perfoliate leaf—a round disk with the stem appearing to grow up through the center. Several basal leaves are present as well, ranging from linear to spade or diamond shaped. The small white flowers have two sepals and five petals. After the plant has been pollinated and the petals have fallen off, the sepals fold back up to protect the fruit as it matures. Miner's-lettuce is abundant in moist shady sites throughout California up to 6,500 feet. The leaf is edible as a salad green and was eaten by the early miners and pioneers, as well as the Native Americans.

Miner's-lettuce

CANDY FLOWER *(Claytonia sibirica)* is an appealing plant, as its name implies, and is in the same genus as miner's-lettuce *(Claytonia perfoliata)*. A rather fleshy perennial, it has larger flowers than miner's lettuce and does not have perfoliate

Candy flower

leaves. The delicate stem can range from only two inches high in open areas to almost three feet in tall dense vegetation where it must compete for sunlight. The stem leaves are lanceolate to ovate and sit opposite each other along the stem, but the basal leaves are narrower. The flowers have five white petals with pink striping and are about half-an-inch long. Found up to 4,000 feet from Fresno and Santa Cruz Counties northward, it occurs in various wet areas such as marshes and streambanks, usually in somewhat shaded areas.

The pink family (Caryophyllaceae) has many showy garden plants, such as carnations (*Dianthus* spp.) and sweet William (*Dianthus barbatus*). Others, such as chickweeds (*Stellaria* spp.), are less showy but can be very common. Plants in this family always have opposite, simple leaves, and the five petals are often deeply divided, sometimes looking as if their edges had been cut with pinking shears. **BEACH STARWORT *(Stellaria littoralis)***, a native chickweed, is a sprawling, hairy, but very attractive and delicate coastal perennial. It has long, wavy hairs and stemless ovate leaves with dense cilia, or hairs, along the edges. The white flowers are about half-an-inch wide with

Beach starwort

petals generally divided about halfway down. This plant has 10 stamens, and the pistil is divided at the top into three tiny, spreading styles. It occurs in marshes and on coastal bluffs and dunes in central and northern California. Once common, beach starwort has become much scarcer in recent years.

Common chickweed

COMMON CHICKWEED *(Stellaria media)* is a small, nonnative ubiquitous plant that can be found in lawns and urban parks and fields as well as in wilder areas. The small ovate leaves are stemless and sit opposite each other on the central stem. The small white flowers have five petals that are so deeply divided they often look like 10 petals at first glance. The distinguishing characteristic of this particular species of chickweed is a narrow line of white hairs running down only one side of the central stem. At each pair of leaves, the line of hairs changes to the next side of the stem, thus appearing to spiral down the stem. This abundant weed is found up to about 4,000 feet in many habitats all over California. It is edible and can be combined with miner's lettuce *(Claytonia perfoliata)* for a tasty spring salad. Be sure to check for the line of hairs on the stem because the plant without its flowers can resemble scarlet pimpernell, which can be poisonous. Less frequent, but more showy and native to California, is field chickweed *(Cerastium arvense)* with larger flowers and elongate rather than spherical fruit. It usually occurs in moist areas and can be found in much of central and northern California, as well as in the Sierra Nevada.

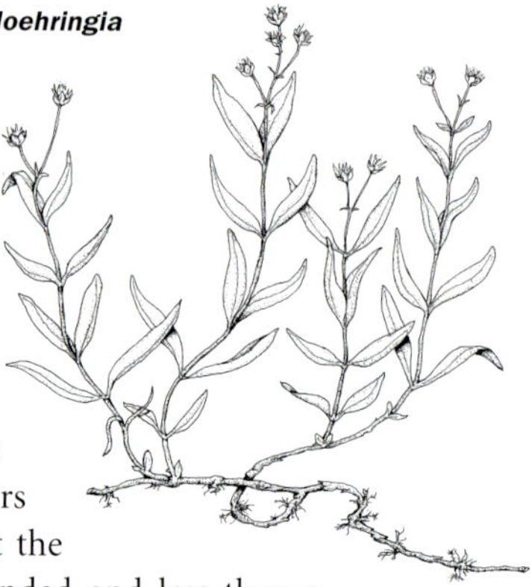

LARGE-LEAVED SANDWORT *(Moehringia macrophylla)* is related to the chickweeds. A perennial growing to about eight inches high, it has an angled or grooved stem with tiny hairs and elliptic to lanceolate leaves that can be up to two inches long and are usually ciliate on the edges. The two to five flowers have five pointed sepals, but the five petals are generally rounded and less than a quarter-inch long. The pistil has three styles, which is often indicative of this family. The plant can be short, dense, and almost matlike, or taller and more open. It is found on shaded slopes between 1,000 and 6,000 feet in the Sierra Nevada, the Coast Ranges the length of the state, and in most of northwestern California to British Columbia.

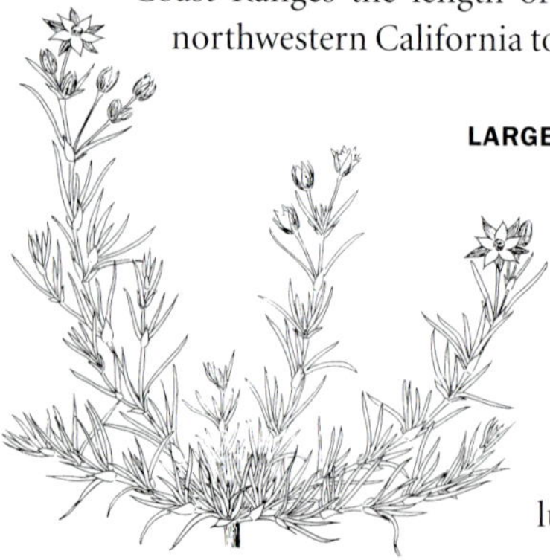

LARGE-FLOWERED SAND-SPURREY *(Spergularia macrotheca)* is a low perennial with fleshy linear leaves bunched together in fascicles (clusters) along a stout stem. At the base of each fascicle is a somewhat papery translucent scale called a stipule.

Large-flowered sand-spurrey

The inflorescence is glandular, and the half-an-inch-wide flowers have five spreading white or rosy pink petals, 10 stamens, and a three-styled pistil. It is found in moist, saline places near the coast or inland through much of California up to 2,500 feet. A related annual species, common sand-spurrey *(S. rubra)*, is smaller and more delicate with pink to lavender flowers. It has a spreading, ground-hugging habit, and is common in drier, usually disturbed sites, often growing in the middle of well-used trails and fire roads.

On beaches and bluffs near the immediate coast is **CRYSTALLINE ICEPLANT** ***(Mesembryanthemum crystallinum)*** in the fig-marigold family (Aizoaceae). It is one of several species of iceplant growing in California that are also popular groundcover plants in landscaping. A prostrate annual from Africa, this is one of the narrower-leaved species with crowded, succulent leaves and white flowers about an inch across that turn pink as they age. The name iceplant refers to the masses of glistening raised epidermal cells that give a sparkling appear-

Crystalline iceplant

ance to the leaves and stems. It blooms from March to October and is common on beaches, dunes, and bluffs along most of the California coast.

The buckwheat family (Polygonaceae), another large family in California, is found primarily in drier areas. The plants usually have simple leaves and stems with swollen nodes where the leaves attach. The small flowers often occur in clusters subtended by an involucre (fused set of usually leaflike bracts below a head, or cluster, of flowers). One of the more widespread members of the family is **CALIFORNIA BUCKWHEAT** ***(Eriogonum fasciculatum),*** a low, semiwoody shrub with linear leathery leaves in fascicles along the stem. The leaf edges are usually rolled under, and the undersides are densely white hairy. The small white to pink flowers are in heads at the tips of the branches, usually with several involucres per head. This species has four varieties that are mainly distinguished by the shape, color, hairiness of the leaves, and degree to which the edges are rolled under. California buckwheat is an important bee plant and occurs in scrub areas on

California buckwheat

dry slopes and in dry washes and canyons throughout the state up to 7,500 feet. Many other species of wild buckwheat *(Eriogonum)*, both annual and perennial, can be found in different parts of California.

In the saxifrage family (Saxifragaceae), the flower forms a tube or cup around the ovary, and the stamens, petals and sepals are attached along its rim. It resembles the rose family in this respect, but most plants in the saxifrage family can be distinguished by their two-parted flower pistils or styles. In addition, most species have a cluster of rounded basal leaves and very few, if any, leaves on the stem. **CALIFORNIA SAXIFRAGE *(Sax-***

ifraga californica) is a typical and very lovely member of this family that blooms early in spring. It has long-stemmed basal leaves with shallow, rounded, or pointed teeth. The flowering stem can be more than a foot tall and has a cluster of several delicate white flowers with 10 stamens with bright red anthers. The two parts of the pistil curve away from each other like the horns on a sheep or goat, which is a feature often seen in the saxifrage family. This species can be found on shaded, often grassy, banks below 4,000 feet in most of California.

UMBRELLA PLANT, or **INDIAN-RHUBARB**, ***(Darmera peltata)*** is a large plant growing along the banks of rapid streams. A leafless flower stem up to five feet tall appears in middle spring with clusters of several white flowers, each of which is less than an inch wide. Later in the season, large basal leaves up to three feet wide appear on long leaf stems that can be more than three feet tall. It grows up to 6,500 feet in the Sierra Nevada, Cascade Range, and Klamath Mountains, and into southwestern Oregon. The young growth was eaten by Native Americans, hence the name Indian-rhubarb.

Umbrella plant, or Indian-rhubarb

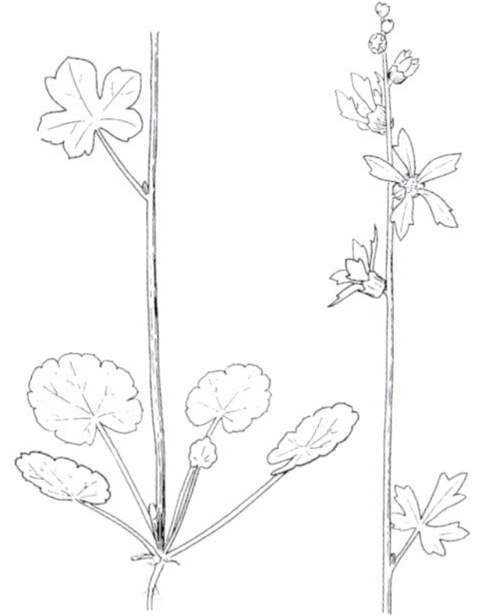

Hill star

In the woodland star genus *(Lithophragma),* the plants have round basal leaves, only a few stem leaves, and white flowers with unevenly divided petals giving a "fringed" appearance. The two most often encountered in California are **HILL STAR *(Lithophragma heterophyllum)*** and **WOODLAND STAR *(L. affine)*.** Hill star is usually shorter, and has smaller flowers that look more "fringed," but the best way to distinguish between the two species is to look at the base of the sepals below the flower. In hill star, the base is U shaped, and in woodland star, it is V shaped. Although their names indicate otherwise, hill star is usually found in moister, shaded areas below 5,000 feet, and woodland star occurs more commonly on open grassy slopes up to 6,500 feet. Both occur in the Coast and Transverse Ranges, but woodland star can also be found in the Peninsular Ranges of southern California, Klamath Mountains of northern California, and central Sierra Nevada. Also in the genus are rock star *(L. glabrum),* mission star *(L. cymbalaria),* and prairie star *(L. parviflorum),* among others.

Plants in the rose family (Rosaceae) range from small herbs such as strawberries (*Fragaria* spp.) to large trees such as apples (*Malus* spp.) and pears (*Pyrus* spp.). The flowers are cuplike, similar to the saxifrage family (Saxifragaceae) described above, but they often have a set of bracts (leaflike structures) below the sepals. Plants in this family also usually have numerous leaves along the stem, unlike the saxifrage family, which has few or none. **HONEY DEW *(Horkelia cuneata)*** is a glandular, herbaceous perennial with flat, open white flowers about half-an-inch wide. The bracts alternate between the sepals, thus giving the impression of 10 sepals, but close inspection reveals that the bracts are a separate layer on the outside. The leaves are divided into 10 to 24 evenly toothed or lobed leaflets. Honey dew is found below 1,500 feet in open, sandy places and in woodlands from San Francisco to San Diego and has three subspecies, one of them quite rare. In central and northern California, as well as the San Bernardino Mountains, a somewhat similar species is California horkelia *(H. californica)*, but the leaflets have wider bases and uneven teeth or lobes and are often palmately veined at the base, whereas in honey dew, the leaflets are pinnately veined throughout. It, also, has three subspecies differing mostly in the number and shape of the leaflets.

BEACH STRAWBERRY *(Fragaria chiloensis)* is one of the parents of domestic strawberries. It has the usual trailing strawberry habit, three leaflets, and white flowers with many yellow stamens. The shiny, leathery leaves are densely hairy below, and the flowers produce a very sweet berry. Strawberries are the only fruit that carry their seeds on the outside rather than protected inside of the fruit. Beach strawberry is found pri-

Beach strawberry

marily in sandy places along the immediate coast from central California to Alaska. An inland species found in shaded areas is wood strawberry *(F. vesca)*, a smaller plant with smaller flowers and fruit.

CALIFORNIA BLACKBERRY ***(Rubus ursinus)*** is a rambling vinelike plant with thick arching stems called canes. It has many slender prickles along these canes and also on the undersides of the three-parted leaves. The white flowers are up to an inch across and produce delicious berries in late spring or early summer. Loganberry, youngberry, and boysenberry have all been developed from this species, which grows in moist shaded places up to 5,000 feet throughout most of California. More common and weedy is the nonnative

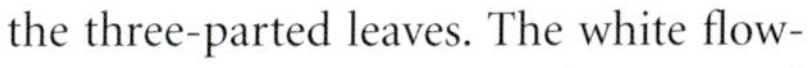

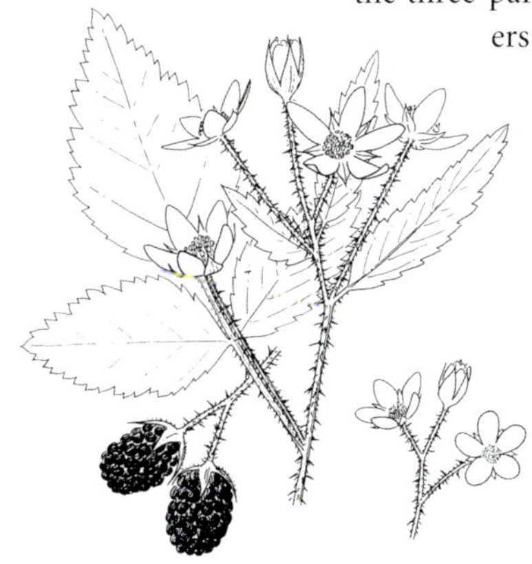

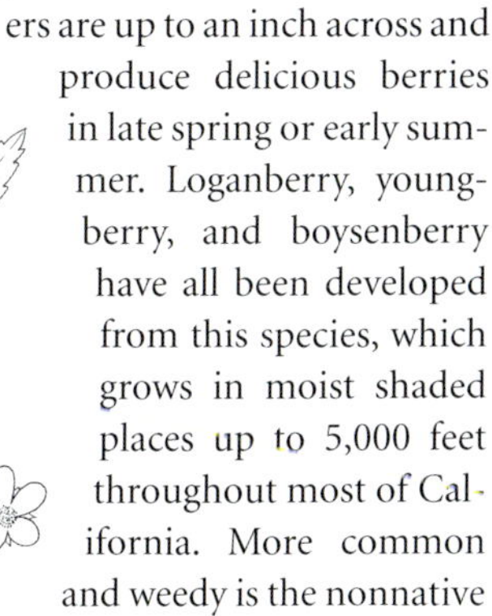

California blackberry

Himalaya blackberry *(R. discolor)*, which has fewer but larger and thicker prickles with broad bases and hooked tips, thus being much more painful when encountered. It flowers and fruits later in the season, the berries appearing about mid-summer or later.

In woods and rocky places near the coast is **OCEANSPRAY**, or **CREAM BUSH**, ***(Holodiscus discolor)***, a slender-stemmed shrub with airy masses of small spiraea-like flowers that range in

Oceanspray, or cream bush

color from white or cream to pale pink. It can be from three to almost 20 feet tall and has thin to leathery leaves toothed only on the upper two-thirds of the margins. Cream bush does not flower for the first several years, but when it matures and begins to produce hanging sprays of flowers, it becomes easily recognizable year-round as the gray, dried-up sprays remain on the plant until the new flowers appear the following spring. It occurs below 6,000 feet on the moist edges of woodlands and on rocky slopes in much of California.

CHAMISE, or **GREASEWOOD**, ***(Adenostoma fasciculatum)*** is one of the chief components of most chaparral areas in California. It is a resinous shrub with shredding red bark and fascicles of stiff linear leaves along the stem. In late spring, dense clusters of minute white flowers appear, each with small rounded petals

Chamise, or greasewood

and clusters of 10 to 15 stamens. The flowers turn a dark rusty brown by midsummer and persist through fall, sometimes coloring vast hillsides. Because of the resin in the branches, chamise is highly flammable and plays a large part in the frequent summer and fall fires that plague southern California. It can be found below 5,000 feet through much of the state on dry slopes and ridges but is especially prevalent in the south.

The pea family (Fabaceae) is easily recognized by its distinctive flower consisting of a single wide upper petal called the banner, two lower wing petals, and a keel petal hidden below the two wings and enclosing 10 fused stamens and a pistil. Within this family, the lupines are one of the most prolific and widespread groups, with more than 70 species in California, all of them native. They grow everywhere from coastal strand to the highest mountain tops, from the sea to the deserts. One of the most common of the white species (although it can sometimes have yellow or pink flowers) is **DENSE-FLOWERED PLATYCARPOS** ***(Lupinus microcarpus*** **var.** ***densiflorus),*** a rather fleshy, hairy annual about a foot tall with flowers whorled around the stem. The leaves are typical of the lupines, having several

Dense-flowered platycarpos

oblanceolate leaflets that spread out palmately (fingerlike) from the tip of a long petiole (leaf stem). When the plant begins to fruit, the two-seeded pods begin to lean over to one side of the stem. The closely related chick lupine *(L. microcarpus* var. *microcarpus)* is very similar, but the sepals are more soft hairy, the flowers are usually pink, and the fruit stays whorled around the stem. Both varieties grow in grassy and open places, often on new road cuts, up to 5,500 feet through most of California.

Spanish clover

SPANISH CLOVER *(Lotus purshianus)* is not a true clover but a member of the *Lotus* genus in the pea family (not to be confused with the Chinese *Lotus* of a different family). It is a prostrate or short annual and the thin stem may be simple or branched. When competing in tall grasses or other vegetation, however, it can become very tall and leggy. Three leaflets occur per stem, and often a pair of very leaflike bracts as well. A single, delicate, pale pink to yellow, pea-shaped flower sits at the end of each flower stem. Although a native, it can become very weedy and is often found in disturbed and waste places and along roadsides, as well as in woodlands and forests, along streams, and in coastal areas up to 8,000 feet throughout California. The flowers appear in middle to late spring, and it can often still be found blooming as late as October.

Some of California's most delicate wildflowers are in the flax family (Linaceae). Although some European species have very tough stems (linen is derived from the outer stem fibers of one species), most of our California species are much more tender. Some have flowers so sensitive that the petals fall off at the slightest touch. The leaves of plants in this family are simple, lanceolate to ovate, and sessile (their base attached directly to the central stalk, i.e., there is no leaf stem), and the flowers have four or five petals and pistils with two to five styles. **CALIFORNIA DWARF FLAX *(Hesperolinon californicum)*** is less than 10 inches tall with very narrow leaves about an inch long arranged alternately along the stem. The small delicate flowers have five white or pink petals, five stamens with white or pink anthers, and a pistil with three styles. It can be found up to 4,500 feet in chaparral or grassland in rocky areas, and sometimes on serpentine, mostly in the inner Coast Ranges but also in the northern Sierra Nevada and the southern Cascade Range. Several species in this family have become quite rare or endangered in California because of loss of habitat and other factors.

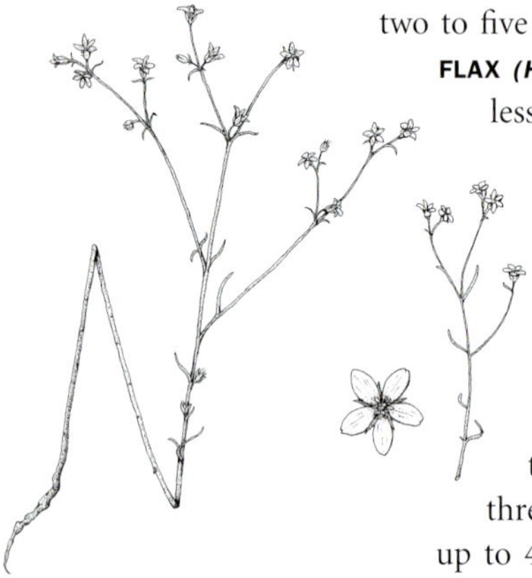

Buckbrush and California-lilac *(Ceanothus)* comprise a large genus of shrubs in the buckthorn family (Rhamnaceae). Buckbrushes have clusters of white or pale blue flowers, horned fruit capsules, and thick leaves with sunken white pits

Bigpod ceanothus

on the underside and arranged opposite each other on the stem, and California-lilacs have clusters of usually deeper blue flowers, fruit capsules without horns, and thinner, three-veined leaves that occur alternately along the stem. **BIGPOD CEANOTHUS *(Ceanothus megacarpus)*** is in the first group, but the leaves are often both opposite and alternate on the same plant. They are dull green, elliptic to obovate, and white hairy on the underside and remain on the plant year-round. The fragrant white or lavender flowers are in short clusters less than an inch long, and the three-parted fruits have obvious horns. The flower is typical of those in this genus, having five erect, hooded petals and five spreading sepals between them, the same color as the petals. In the center of the flower is a flat, round disk with the stamens attached along the edge. This plant occurs below 3,000 feet on dry, shrubby slopes and in canyons near the coast from central California southward and in the Channel Islands. Closely related species vary in leaf shape, degree and type of notching at the leaf tip, and shape and size of the horns and ridges on the fruit capsule.

Salal

SALAL *(Gaultheria shallon),* in the heath family (Ericaceae), is related to the native manzanita, madrone, and huckleberry, as well as the wintergreen of East Coast woods. The heath family has urn shaped flowers and simple, often leathery, leaves. Salal is usually a low, spreading shrub, but can sometimes be up to five feet tall. It has tough evergreen leaves two to four inches long, and clusters of white or pink flowers with petals fused into an urn or bell shape. The fruit is a dark purple capsule with several seeds inside. It is found up to almost 3,000 feet in moist forest margins near the coast, mostly in northern California, but less frequently as far south as Santa Barbara. The thick leaves are known as lemon leaves in the florist trade and have often been used on floats in the famous Tournament of Roses Parade held on New Year's Day in Pasadena.

Related to salal *(Gaultheria shallon)* and with the same urn-shaped flowers is **GREATBERRY MANZANITA *(Arctostaphylos glauca)***. It belongs to one of California's most interesting and variable genera with more than 40 species in the state, as well as many hybrids. Varying from erect and almost treelike species in the interior to prostrate plants such as bearberry *(A.*

Greatberry manzanita

uva-ursi) on the north coast, manzanitas have smooth, reddish bark and petite pink or white urn-shaped flowers. The small, round, reddish fruit gives the genus its common name, manzanita, which means "little apple" in Spanish. Greatberry manzanita has thick, rounded, oval to oblong gray green leaves and can reach a height of 25 feet. The urn-shaped flowers are white to pink with short, glandular, bristly stems. The seeds are fused into one large stone inside the fruit, whereas the fruit of most other manzanitas contain several seeds. This species is found on rocky slopes in chaparral up to 4,500 feet from the San Francisco Bay Area southward. In northern California and the Sierra Nevada, a frequently encountered manzanita is common manzanita *(A. manzanita)*, with smaller, glossier leaves. Manzanitas hybridize freely, and it is often difficult to distinguish between various species and hybrids, the characteristics of two or more species sometimes being apparent on the same plant.

SHAGGY-BARKED MANZANITA *(Arctostaphylos tomentosa)* is another erect shrub, but it only reaches a maximum height of eight feet and is usually even shorter. It has dark or bright

Shaggy-barked manzanita

green leaves that are somewhat shiny on the upper surface and usually duller and hairy on the underside and twigs that are generally densely hairy or feltlike and often glandular as well. The flowers are in crowded clusters and have densely hairy and sometimes glandular or bristly stems and densely hairy ovaries. Nine subspecies exist, some of them with different common names, differing mostly in leaf and twig characteristics. One or more of the subspecies occur up to 4,000 feet from the San Francisco Bay Area southward, and two are found only in the Channel Islands. Manzanitas are a main component of chaparral in California, and there are few brushy chaparral slopes in the state without one form or another of manzanita. The fruits of some species, if not too dry, make an excellent jelly.

WESTERN AZALEA *(Rhododendron occidentale)* is another member of the heath family. A loosely branched deciduous shrub three to 14 feet tall, it has light green elliptic to oblong leaves one to three inches long with small hairs along the edges. The

Western azalea

large, showy flowers are white or pink, often with a yellow blotch at the base of the upper petal. They are funnel shaped below, spreading into five large petal lobes above, and have five very long stamens in the center of the flower. It is found in damp places and along streams, mostly in coniferous forests below 7,000 feet in the pine belt in the mountains of San Diego and Riverside Counties and at lower elevations in the Coast Ranges, Sierra Nevada, and northward. It flowers from April to August and is reported to be poisonous to stock.

SNOWDROP BUSH ***(Styrax officinalis*** **var.** ***redivivus)*** is the only California member of the storax family (Styracaceae). A sweet-smelling bush from three to 13 feet tall with pendulous white flowers, it has simple, alternate, deciduous leaves that are generally glabrous (not hairy) on the darker upper surface but stellate haired (having multibranched hairs, often star shaped) on the paler green underside. The flowers are about an

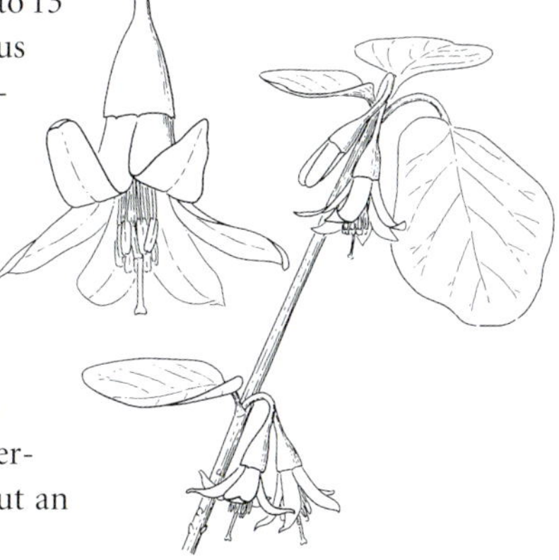

inch long or less, with the petals fused into a tube below but separating into four to 10 lobes above. It is occasional in canyons and rocky places among chaparral shrubs or in woods below 5,000 feet in many parts of California.

The morning-glory family (Convolvulaceae) is familiar to most people with its colorful funnel-shaped blossoms and climbing stems. **WESTERN MORNING-GLORY *(Calystegia occidentalis)*** is a native species with white or creamy yellow flowers one to two inches long. The plant is at least slightly hairy, especially on the leaf or flower stem, and can sometimes be almost feltlike. The triangular leaves are about an inch long and

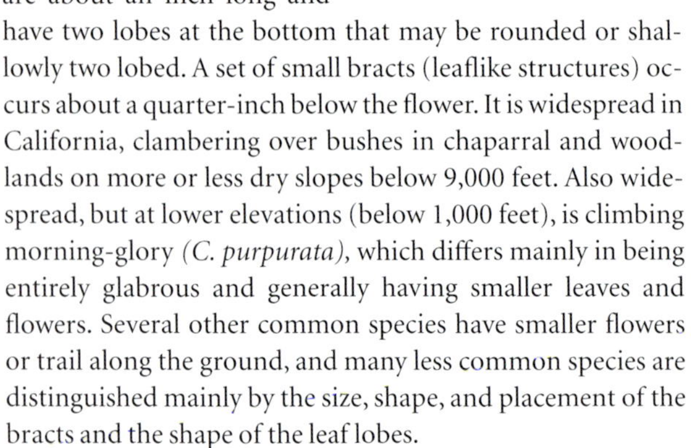

have two lobes at the bottom that may be rounded or shallowly two lobed. A set of small bracts (leaflike structures) occurs about a quarter-inch below the flower. It is widespread in California, clambering over bushes in chaparral and woodlands on more or less dry slopes below 9,000 feet. Also widespread, but at lower elevations (below 1,000 feet), is climbing morning-glory *(C. purpurata)*, which differs mainly in being entirely glabrous and generally having smaller leaves and flowers. Several other common species have smaller flowers or trail along the ground, and many less common species are distinguished mainly by the size, shape, and placement of the bracts and the shape of the leaf lobes.

Plants in the evening-primrose family (Onagraceae) have flowers with four petals and long, thin inferior ovaries (attached below the petals and sepals) that sometimes look like part of the stem. The family gets its name from its many species that don't open until sundown and close up the next

Devil's lantern, or lion-in-a-cage, or basket evening-primrose

day when the sun gets bright. A white-flowered species is **DEVIL'S LANTERN** ***(Oenothera deltoides)***, also known as **LION-IN-A-CAGE**, or **BASKET EVENING-PRIMROSE.** A grayish annual or perennial, it has a loose rosette of leaves at the base and large, coarsely toothed or coarsely lobed leaves arranged alternately along the stem. The large flowers are in the upper axils (angle formed where the leaf stem meets the main stem). The petals can be more than an inch long, and the stigma at the top of the pistil has four linear lobes. Four subspecies are found in California, varying as to type of hair and shape of buds. They occur below 6,000 feet, mostly in sandy desert areas and can form large showy masses, but one, Antioch Dunes evening-primrose *(O. deltoides* subsp. *howellii)*, is rare and endangered, occurring only in the Antioch Dunes area near the Delta east of San Francisco Bay.

The phlox family (Polemoniaceae) has a number of small, graceful annual wildflowers in California, such as **BICOLORED LINANTHUS** ***(Linanthus bicolor)***. Flowers in this family have ribbed sepals with thin, translucent connecting membranes between them; flower tubes that expand into a throat at the

Bicolored linanthus

top and then spread into five petal lobes; and pistils with three-parted stigmas. Bicolored linanthus has opposite leaves with thin, narrow lobes and a head of several flowers, only one or two of which are usually in bloom at a time. The flower tube, about an inch long or less, is reddish, the throat is yellow, and the spreading, rounded petal lobes are white or pink. The stigma is less than one millimeter long. It is common and often plentiful in grasslands, woodlands, chaparral, and many other habitats below 6,000 feet in the Coast Ranges, Sierra Nevada, western Transverse Ranges, Channel Islands, and San Francisco Bay Area. It is very similar to another common species, small-flowered linanthus *(L. parviflorus)*, which has a longer, thinner flower tube. The petal lobes are larger than in bicolored linanthus, in spite of its common and scientific names, and the stigma is three to six millimeters long.

Another type of linanthus much more open and branched is **FLAX-FLOWERED LINANTHUS *(Linanthus liniflorus)*.** It grows to almost two feet in height, and the leaves are deeply lobed into linear segments. The plant is diffusely branched, and the

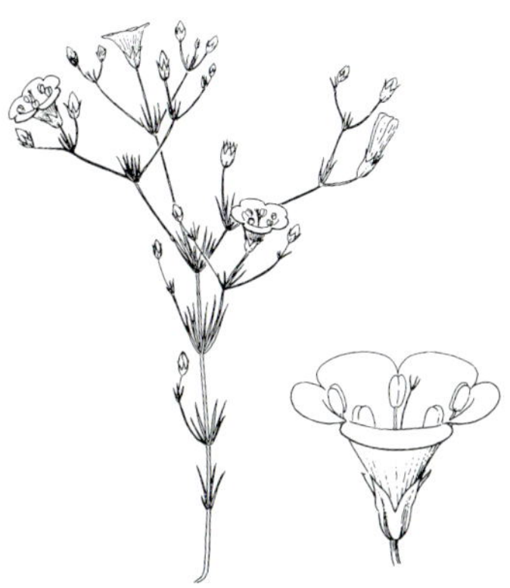

Flax-flowered linanthus

flowers are on thin, threadlike pedicels (flower stems). The flower tube is very short, and the wide throat is longer than the tube. The white petal lobes are about half-an-inch long or less and have conspicuous purple veins. It is found in woodlands, deserts, and other open dry places, often on serpentine, below 5,000 feet from southern California to Washington. Many other species of this delicate flower are found throughout the state.

A more or less spiny plant in the phlox family is **WHITE-FLOWERED NAVARRETIA *(Navarretia leucocephala)*.** It usually does not develop a central stem but forms flat masses on the ground. The leaves are divided into thin lobes that may themselves be divided again. Below the head of flowers are several bracts looking very much like the leaves, but a closer look reveals that that they are of a different design with broad bases and two small, sharp spines at the tips. The tiny white flowers are less than half-an-inch long and have spine-tipped sepals. It is found in the dried mud of vernal pools (depressions that have held water during the rainy season). It occurs in most of northern California below 7,000 feet and to Washington and Nevada. Five subspecies exist, three of them being very geographically limited and designated as rare and endangered.

Plants in the waterleaf family (Hydrophyllaceae) often have divided leaves and flowers in coiled clusters. **YERBA SANTA *(Eriodictyon californicum)*** is an exception with entire, simple leaves and clusters of flowers that are not obviously coiled. It is also one of the few woody members of the family, being an aromatic shrub that can be five or six feet tall. It usually has entire, glandular leaves up to six inches long and about two inches wide. The funnel-shaped flowers are bluish lavender to white and one-fourth to one-half inch long. It ranges through most of northern California up to 6,000 feet and into Oregon. In southern California it is replaced by hairy yerba santa *(E. trichocalyx)* with somewhat smaller, more hairy flowers. In the early days yerba santa was used to make a tea for colds and asthma.

FIVESPOT *(Nemophila maculata)* is in another genus of the waterleaf family that does not have its flowers in coiled clusters. Rather, it can be recognized as a member of this family by its small reflexed sepal-like appendages between the sepals. A

Fivespot

low, spreading annual with opposite, lobed leaves, fivespot has bowl-shaped white flowers about an inch across or more with a distinctive deep purple blotch at the tip of each petal. It occurs in meadows and woodlands below 10,000 feet in the Sierra Nevada and the Sacramento Valley. Other species of *Nemophila* may be white or various shades of blue.

Another family with coiled flower stalks is the borage family (Boraganaceae), but the flower stalks are solitary or in loose, not tight, clusters. The flower petals are fused into a short tube below and spread out above into five petal lobes that usually have a small appendage at their base, often forming a crown at the center of the flower. Forget-me-nots *(Myosotis latifolia)* and the deep blue garden borage *(Borago officinalis)* are in this family, and **SEASIDE HELIOTROPE** ***(Heliotropium curassavicum)*** is a native perennial member. Prostrate or ascending, it has fleshy oblanceolate leaves and many small identical, bell-shaped, white to bluish flowers, often purplish in the center, spaced symmetrically on a fleshy arching stem. It can be found in somewhat saline, often sandy, places below 7,000 feet throughout California. (See photograph, page 72.)

Seaside heliotrope

Many species of popcornflower occur in California, but the most common is one that is simply called **POPCORNFLOWER** ***(Plagiobothrys nothofulvus)***. It has a basal rosette of leaves, mostly alternate stem leaves, and coiled flower stalks with brown- or tan-haired buds when young, becoming long and arched as the flowers open, mature, and go to seed. The small white flowers are circumscissile, that is, the upper petal lobes come off as a round unit rather than petal by petal. The lower part of the plant has a purplish dye that might come off on your fingers when handled. It is found in relatively dry, open grasslands and woods below 3,000 feet throughout California. The many other species of popcornflower all look very similar and can often only be distinguished from one another by examining their mature fruit under a microscope.

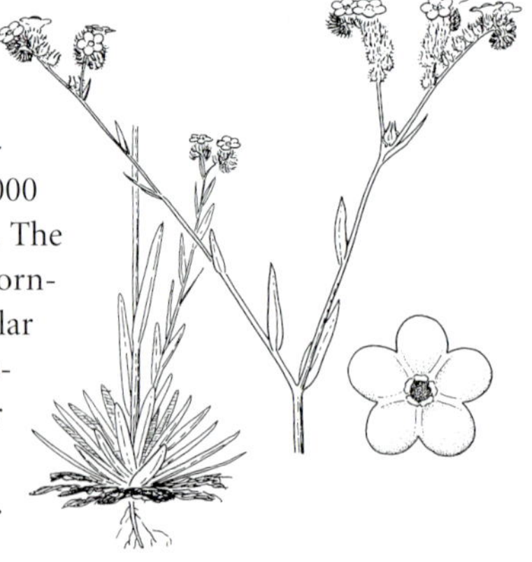

Popcornflower

They all have prostrate to erect stems and coiled flower stalks with several small white flowers, but some species have flowers near the base, whereas others don't; some have alternate leaves near the base, some have opposite; some have basal rosettes of leaves, some don't. Several species occur in wet places, others in drier areas. One species or another can be found almost anywhere in the state.

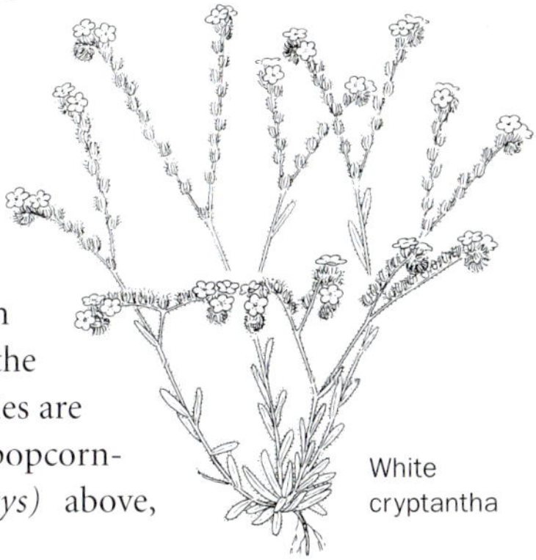

Related to popcornflower, but generally occurring in drier habitats and having stiffer hairs, no purple dye, and smaller flowers with mostly smooth seeds is the *Cryptantha* genus. The species are hard to tell apart, as in the popcornflower genus *(Plagiobothrys)* above,

White cryptantha

differing mostly in fruit, flower size, and type of hairs. **WHITE CRYPTANTHA** ***(Cryptantha intermedia)*** is a rough, bristly-haired annual, usually with ascending branches of several white flowers with very rough-hairy sepals and petal lobes only a quarter-inch wide or less, usually with a bright yellow appendage at their base. The fruit often consists of only one nutlet per flower, rather than the more typical two to four of the family, and is rough rather than smooth. It is found from 1,000 to 9,000 feet in sandy to rocky soils in oak woodlands and coniferous forests from Baja California to British Columbia. Other *Cryptantha* species occur in dry, open places, such as chaparral burns, through much of the state.

The mint family (Lamiaceae) is noted for its many fragrant members with essential oils, such as bergamont, thyme, pennyroyal, catnip, and mint itself. The stems are square, the leaves are opposite each other, and the flowers are usually tubular below and spreading into two lips above: a two-lobed upper lip and a three-lobed lower lip with the middle lobe generally extending tonguelike beyond the two side lobes. Among our natives are several interesting species of sage, such as **WHITE SAGE** ***(Salvia apiana)***, an important bee plant, as indicated by its scientific name. Shrubby below with velvety, white, lanceolate leaves, it has a flower stalk that reaches a height of three feet or more with openly branched spikelike clusters of white flowers tinged with purple. It grows on dry slopes in coastal-sage scrub, chaparral, and yellow-pine forests up to 5,000 feet from Santa Barbara County to Baja California.

White sage

A plant having a fragrance more delicate than the preceding white sage and usually growing flat on the ground is **YERBA BUENA *(Satureja douglasii),*** meaning "good herb" in Spanish. With slender, trailing stems and oval leaves up to about an inch long with round-toothed edges, it bears one to three small white, mint-shaped flowers in the leaf axils (angle formed where the leaf stem connects to the main stem). Found in shady places below 3,000 feet in chaparral, coastal scrub, and woodlands, it occurs from central California northward and in the Transverse Ranges of southern California. The odor of the crushed foliage is very pleasant, and it was used by Native Americans and the early settlers as a tea for drinking and as a medicine for fevers and stomach complaints.

California man-root, or California wild-cucumber

CALIFORNIA MAN-ROOT, or **CALIFORNIA WILD-CUCUMBER**, ***(Marah fabaceus)*** is an interesting native vine in the gourd family (Cucurbitaceae), a family with plants having trailing or climbing stems with tendrils, separate male and female flowers, and gourdlike or melonlike fruit capsules or berries. California man-root has wide, lobed leaves on masses of long climbing vines 15 to 20 feet high that clamber over bushes and trees, sometimes almost smothering them. It has white, creamy, or yellowish green flowers, and the several male flowers are on single flower stems rising from the leaf axils, whereas the single, stemless female flower sits below, eventually producing a spiny round fruit about two inches long with five marble-size brownish seeds. The above-ground part of the plant dies back each summer, and the voluminous climbing growth is produced anew each year. The remarkable root can be up to five feet long, the size of a small man. It is shaped like a huge carrot or beet, weighs many pounds, and lasts from year to year, storing the great amounts of water and starch needed to produce the impressive fresh growth each year. It can be found in shrubby and open areas and near streams and washes up to about 5,000 feet throughout California.

The honeysuckle family (Caprifoliaceae) has opposite, usually simple leaves and flowers with inferior ovaries (attached below the sepals and petals). **CREEPING SNOWBERRY**, or **TRIP VINE**, ***(Symphoricarpos mollis)*** is a low, partly trailing shrub in this family with thin branches and frail rounded leaves. The two to eight light pink flowers per cluster are bell shaped and about a quarter-inch long. The round inferior ovary becomes a white berrylike fruit twice the size of the flower. It is com-

Creeping snowberry, or trip vine

mon on shaded slopes below 10,000 feet throughout most of California. Also common is a taller, more erect shrub, known simply as **SNOWBERRY** *(S. albus* var. *laevigatus)*, with eight to 16 flowers per cluster, each flower somewhat swollen on one side. In both species the flower is hairy on the inside, distinguishing them from other species of snowberry.

A small native relative of the popular garden and landscaping plant red valerian *(Centranthus ruber)* is **WHITE PLECTRITIS** ***(Plectritis macrocera)*** in the valerian family (Valerianaceae). The two plants bear little resemblance to each other until you look more closely at the individual flowers. Both have clusters of very similar two-lipped tubular flowers with an erect upper

White plectritis

lip and a spur protruding from near the base of the tube. White plectritis is a much smaller and more delicate plant with fewer flowers per cluster. Rarely over a foot tall, it has opposite, more or less spoon-shaped leaves with delicate network veining that can sometimes be reminiscent of stained glass. The flowers are white or pale pink and less than a quarter-inch long, and the spur is short, thick, and blunt. It is found throughout California below 6,500 feet on grassy and wooded slopes and often occurs in large numbers. Other species are also widespread. They are usually only a few inches tall and vary in the length and thickness of the spur and the size and color of the flower.

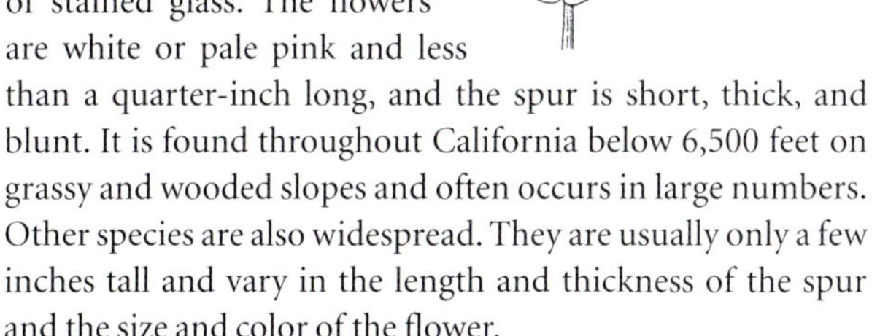

One of the largest families in California is the sunflower family (Asteraceae). The flowers are in heads that look like a single flower but are actually composed of many small individual flowers that may be either disk flowers (small and tubular with five petal lobes) or ray flowers (long, thin, and resembling a single petal), or the head may contain both as in the common sunflower or daisy. Under each head of flowers are several overlapping bracts, or leaflike structures, called

Coltsfoot

Coltsfoot

phyllaries. The family is so large that it is often divided into tribes differentiated, in part, by whether the flower heads have disk flowers, ray flowers, or both, and by the size, shape, and number of phyllaries. In **COLTSFOOT** ***(Petasites frigidus* var. *palmatus)*** the flower heads are composed of mostly disk flowers, although a few ray flowers may be present. It is a perennial herb with thick, creeping rootstocks that send up mostly leafless stems about two feet tall in early spring topped by dense, terminal clusters of white to pale pink, mostly disk-flowered heads. Male and female flowers are usually on separate plants. The large, round leaves don't appear until later in the season and have very long stems that arise from the base of the plant. Coltsfoot is found below 1,500 feet in deep shade, usually in streams or very wet soil, in the Coast Ranges from the Santa Lucia Mountains northward to Alaska.

WHITE LAYIA ***(Layia glandulosa)*** is one of the members of the sunflower family that has both disk and ray flowers. It has a daisylike flower head made up of yellow disk flowers in the center and white or yellow ray flowers around the outside. The plant is quite glandular, as the scientific name implies, and the leaves are narrow with the lower ones usually irregularly lobed. The flower heads are on stems that can be up to three inches long, and the phyllaries below the heads are covered with cobwebby hairs. It occurs in open sandy soils, sometimes in dense masses, up to 9,000 feet. Its range is from

White layia

central western California southward, as well as eastward through the deserts to New Mexico. It can also be found in the Sierra Nevada and Cascade Range, as well as in Washington, Idaho, and Utah. Several species of layia occur in California, many of them resembling each other and requiring close examination of the disk flowers for technical differences to tell them apart. Several, however, are limited to certain soil types, which can be helpful in determining the species at hand.

Yarrow, or milfoil

YARROW, or **MILFOIL**, ***(Achillea millefolium)*** is an aromatic perennial herb in the sunflower family with finely dissected leaves and flat clusters of small white flower heads composed of several disk flowers in the center and three to eight rounded ray flowers around the outside. At first glance, the individual heads can look very much like flowers from another family, especially when the disk flowers are still in bud—the rounded ray flowers appear very petal-like, and the disk flowers in bud are stamenlike. A closer look, however, reveals the phyllaries below each head and the pistils protruding from the ray flowers, confirming it as a sunflower family member. It is found around the world in many habitats and occurs up to 11,500 feet in California. "Yarrow" is an old English name, and "milfoil" describes the numerous segments of the finely divided leaf. The scientific name refers to the ancient warrior Achilles, who first discovered the healing nature of the leaves, which can be steeped in hot water to stop the flow of blood in cuts and wounds. The plant has many medicinal uses and is a very strong herb.

REDDISH FLOWERS

Rose to Purplish Red or Brown

Checker-lily, or mission bells

In England, several native California species have become popular garden plants, such as the fritillary in the lily family (Liliaceae). One of the more common native fritillaries in California is **CHECKER-LILY**, or **MISSION BELLS**, ***(Fritillaria affinis)***. It has the typical long, thin leaves of the lily family, but the lower leaves are whorled around the stem, and the upper are alternate. The nodding, cup-shaped flower, more than an inch long, is striking, although it can easily blend into the landscape and be missed because of its strange coloring—brownish purple with yellow or yellow green mottling. It grows from a bulb that produces only broad, juvenile leaves on the ground for the first two or three years. As these wide leaves photosynthesize, they produce nutrients that are stored in the bulb until enough have accumulated to provide the strength needed to produce a flowering stalk in the third or fourth year. Checker lily is found in woodlands, scrub, and grasslands below 6,000 feet from the San Francisco Bay Area northward to British Columbia. Other species have flowers with similar coloring, but some have white, scarlet, or orange flowers.

BROWNIES, or **CALIFORNIA FETID ADDER'S-TONGUE**, ***(Scoliopus bigelovii)***, also known as **SLINK POD**, is a strange, but interesting plant. Its short, narrow rhizomes (underground stems) produce two mottled, basal leaves and a pretty but rancid-smelling flower. Pollinated by flies, the flower gives off a smell of rotting meat, and it is difficult to examine the flower closely while holding one's breath. The three broad, spreading sepals and

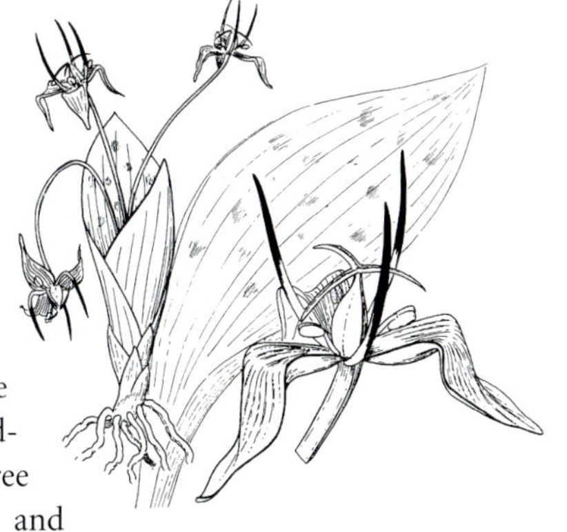

Brownies, or California fetid adder's-tongue, or slink pod

three narrow, erect petals are green or yellow green with purple or brown mottling, striping, or both. Each flower has a very long stem, up to eight inches long, and once the flower starts going to seed, the stem arches over to the ground, thus planting the seed. Occurring up to 3,500 feet in moist, shady areas in redwood forests from Santa Cruz County northward, it is a very early bloomer (January or February) and has usually seeded itself by March.

A large, beautiful plant found in damp redwood forests is **RED CLINTONIA *(Clintonia andrewsiana)*.** Growing from rhizomes, it produces a leafless stem arising from a cluster of five or six wide basal leaves six to 12 inches long and topped by several reddish or rose purple flowers with expanded, pouchlike bases. The flowers are cylindric to bell-shaped, up to two

Red clintonia

inches long, and produce beautiful dark blue berries. It is scattered in redwood and other coniferous forests below 1,500 feet from central California northward to southwest Canada and is a very striking plant to encounter. It blooms from May to July.

Giant trillium, or giant wakerobin

GIANT TRILLIUM, or **GIANT WAKEROBIN**, ***(Trillium chloropetalum)*** is another impressive spring flower and an early bloomer. A much larger plant than the western trillium *(T. ovatum)* (See "Whitish Flowers"), it has the same whorl of three leaves with network veining rather than the usual parallel veining pattern found in the lily family, but the leaves are much longer, broader, and usually mottled. The flower rises directly from the center of the whorl of leaves, lacking the thin flower stem of the smaller species. The flower itself is larger with erect sepals and petals, and the color ranges from pale pink to deep burgundy, or it can occasionally be white or yellow green. It is found on moist slopes and canyon banks in woodlands, forests, and chaparral below 6,500 feet from the San Francisco Bay Area northward to Washington, and it blooms from March to May.

Fifty species of wild onions occur in California, most of them native. **SERRATED ONION *(Allium serra)*** is one of the more common species and grows to about 16 inches tall with a tight, rounded cluster of 10 to 40 small pink to rose flowers with identical petals and sepals that taper to a point. It has two to three long, thin, more or less cylindric leaves at the base of the stem. The scientific and common names refer to the toothed or serrated pattern on the scaly covering, or onion skin, of the bulb. It occurs on grassy slopes below 4,000 feet in the Coast Ranges the length of the state, blooming from March to May. Many other species of onion look similar and can only be identified by examining the patterns on the bulb covering, which, unfortunately, involves digging up, and thus killing, the plant (a practice that is strongly discouraged). Other differences are based on the shape of the ovary and whether or not the ovary has crests (ridgelike bulges at the top). The number and shape of the leaves, which are always basal, and the color, shape, and size of the petals and sepals are also important.

Serrated onion

SCYTHE-LEAVED ONION *(Allium falcifolium)* is another relatively common onion and easily distinguishable from the above species. It has a much lower growth habit and broader, flat,

Scythe-leaved onion

sickle-shaped leaves. The flower stem is short, being less than eight inches tall, and the flower clusters are fewer flowered and not rounded. The flowers are rose purple to white, and some of the petals are usually toothed and have minute glands. The ovary has three low, wide crests. It grows in heavy or rocky soil, such as gravel or talus, and often on serpentine outcrops below 7,000 feet from Santa Cruz County to southern Oregon. It flowers from March to July.

Firecracker flower

FIRECRACKER FLOWER *(Dichelostemma ida-maia)* is a final reddish member of the lily family. It grows to a height of one to almost three feet and has glaucous basal leaves keeled down the center. The several nodding, bright red flowers are in open clusters and look very different from most species in the *Dichelostemma* genus, which usually have tight clusters of small blue flowers. In this species the flowers are tubular and narrow and about an inch long and have small green petal lobes at the tip. Three broad white stamens inside the flower alternate with three white or yellowish false stamens. It is found up to 6,500 feet in canyons, forest edges, and open grassy areas, usually near the coast, from the northern San Francisco Bay Area to southern Oregon, and begins to bloom in May.

STREAM ORCHID ***(Epipactis gigantea),*** a widespread member of the orchid family (Orchidaceae), has a short creeping rootstock and leafy stems from one to two feet tall. The three to 15 flowers are yellowish to green to reddish with red purple veins and usually occur only on one side of the stem. The lower, lip-

Stream orchid

like petal is deeply concave on the lower half and grooved to the tip on the upper half. Growing in moist places below 7,500 feet, it occurs throughout most of California, flowering from May to August. Somewhat similar, but with smaller flowers and growing in dry places, is European helleborine *(E. helleborine)*, a weedy, nonnative species with white to greenish petals and sepals tinged with pink. The lower lip petal is flat above the middle.

California peony

California has two native species in the peony family (Paeoniaceae) that are very different from the popular garden plant with dense pink and white multipetaled flowers. **CALIFORNIA PEONY *(Paeonia californica)*** is the larger of the two natives, with stems up to almost three feet tall. The somewhat fleshy leaves are deeply divided into narrow segments, dark green above and paler green on the underside. The flowers have numerous stamens and somewhat narrow dark red petals edged in pink, one-half to one inch long and larger than the inner sepals. It occurs in chaparral and coastal scrub below 5,000 feet from Monterey County southward. Western peony *(P. brownii)*, the other native, is a shorter plant, also with fleshy divided leaves, but with broader, more rounded segments. The petals are wider and more rounded, less than half-an-inch long, and shorter than the inner sepals. They are maroon to bronze with yellow or greenish edges. It can be found in scrub and open, dry pine forests up to 10,000 feet in the Sierra Nevada and from the San Francisco Bay Area northward to British Columbia.

In the buttercup family (Ranunculaceae), closely related to the peony family (Paeoniaceae), is another garden favorite, columbine *(Aquilegia)*. A common native red species is **NORTHWEST CRIMSON COLUMBINE *(Aquilegia formosa)*,** with leaves deeply lobed to dissected into several segments, looking a little like a bird's claw (another name for the family is crowsfoot because of the leaf shape of many species in the family). The basal and lower leaves are long stemmed, but the upper leaves are on much shorter stems or stemless. The flowers are in nodding clusters and have five spreading red sepals. The five unusual red petals are each rounded into a pointed, horizontal tube with a short straight or incurved spur at the base and a yellow petal lobe at the top. Many species of columbine occur throughout the northern hemisphere, but only three are found in California. This species is the only one that is common in most parts

Northwest crimson columbine

of the state up to 11,000 feet, being found in moist to wet areas in several different habitats. A similar species that can be locally common in the Coast Ranges is Van Houtte's columbine *(A. eximia)*, which differs in having no petal lobe at the top of the tube and a spur that is outcurved. It is often found on serpentine soils but occurs on other moist soils as well, mostly in mixed-evergreen and coniferous forest. The third species, Coville's columbine *(A. pubescens)*, has pale pink to cream-colored flowers and occurs only at high elevations in the Sierra Nevada.

The color blue is usually associated with larkspurs, also in the buttercup family, but one of the few red species is **CARDINAL LARKSPUR**, or **SCARLET LARKSPUR**, ***(Delphinium cardinale)***. Growing from one to more than six feet in height, it has large, deeply dissected basal leaves that are usually withered by the time the flowers bloom and somewhat smaller leaves along the stem. The flower is another uniquely shaped one, with five large showy, petal-like sepals, the upper having long spurs. In the center are four much smaller and inconspicuous petals. It occurs from Monterey County southward in chaparral on dry, often rocky slopes between 1,000 and 5,000 feet. A similar, but usually smaller species, red larkspur, or orange larkspur, *(D. nudicaule)*, is found in the northern part of the state but in somewhat moister conditions on wooded, rocky slopes up to 8,500 feet.

Cardinal larkspur, or scarlet larkspur

California pitcher plant, or California cobra-lily

Another oddly shaped plant is **CALIFORNIA PITCHER PLANT**, or **CALIFORNIA COBRA-LILY**, ***(Darlingtonia californica),*** in the pitcher plant family (Sarraceniaceae). It is a carnivorous perennial, trapping and digesting insects with its long, tubular leaves that flare into arched hoods with two yellow, purple, or green lobes hanging off the tips. The flower, on a long stalk up to about three feet high, has five yellow green sepals with purple veins, and five shorter, narrower, dark purple petals. It is found in seeps and bogs below 7,000 feet in the Klamath Mountains and Cascade Range of northern California and in the northern Sierra Nevada.

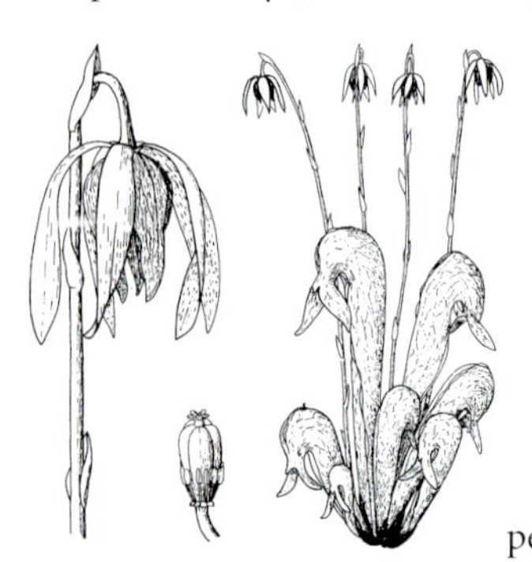

FIRE POPPY ***(Papaver californicum)*** is a red-flowered member of the poppy family (Papaveraceae). Often found after fires, it is

Fire poppy

a slender annual, one to two feet tall with a milky sap and several deeply divided leaves about one to three-and-a-half inches long. The orange to brick red, four-petaled flowers, about an inch wide, have two sepals that fall off immediately after the flower opens. The seedpod is short and flattened at the top. Fire poppy occurs on burns and in recently disturbed places below 1,500 feet from Marin to San Diego Counties. It is very similar to wind poppy, which has orange petals and clear sap. The two species often occur together after a fire.

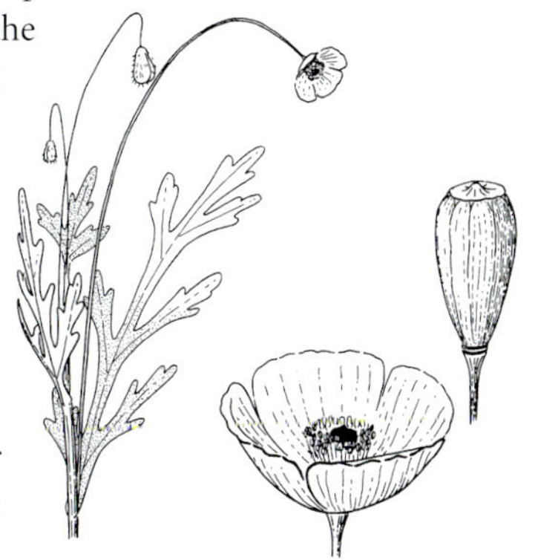

Coast rock-cress

Brewer's rock-cress

A large genus in the mustard family (Brassicaceae) is rock-cress *(Arabis)*, and there are several representatives in California. **COAST ROCK-CRESS *(Arabis blepharophylla)***, growing from a woody caudex, has a few-leaved stem up to about a foot high and a rosette of basal leaves. The stem and leaves are often covered with small, branched, starlike hairs. The flowers, a beautiful shade of rose purple, have four spoon-shaped petals and sepals with bulging, saclike bases. The erect seedpods are long, thin, and flat. It is found in rocky or grassy areas below 1,500 feet along the coast from Sonoma to Santa Cruz Counties.

BREWER'S ROCK-CRESS *(Arabis breweri)* also has beautiful rose purple flowers but is usually found more inland and at higher elevations. Generally more branched and densely hairy than the above species, it has narrower basal leaves that may be clustered but are not in strict rosettes. It also has longer and more numerous stem leaves, though still not a great many, and the sepals do not have saclike bases. The fruit can be erect or spreading, and the petals range from pink to almost purple. It is found in rocky habitats in much of the state, usually above 1,000 feet. It blooms very early and can sometimes be found in flower by November in years of early rains.

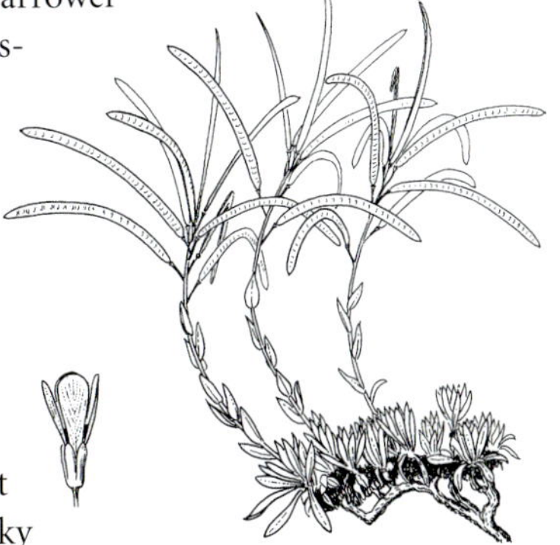

In the mountains is a third species of rock-cress with purple flowers, **FEW-FLOWERED ROCK-CRESS *(Arabis sparsiflora)***. It is generally a taller plant than the above two species, with stouter, gray stems, a cluster of several grayish basal leaves, and many clasping stem leaves that are often sagittate (arrow shaped with two basal lobes). The flowers range from pink to

Few-flowered rock-cress

purple and are about one-half to one inch long. The long, thin seedpods can be erect to recurved. It is found on rocky slopes and in valleys up to 9,000 feet in the Cascade Range, Sierra Nevada, Great Basin, and much of southwestern California.

JEWELFLOWER *(Streptanthus glandulosus)* is a charming member of the mustard family that often occurs on serpentine. Two rare subspecies are found in California, as well as the more common subspecies, *S. g.* subsp. *glandulosus,* which is illustrated. It is a highly variable plant, and continuing research may eventually reveal that there are several more subspecies. An annual, it is often bristly on the lower part of the stem and can grow up to

Jewelflower

three feet tall. The scientific name is Greek for "twisted flower," referring to the four very wavy-edged petals. The sepals are erect and form an urn-shaped cup. In this subspecies they are deep purple, and the petals are a lighter shade of purple. In other forms and subspecies the flowers often occur on only one side of the stem; the sepals can be white, yellow green, or varying shades of rose to red or purple; and the petals range from white to purple. The fruit in all of the subspecies is long and thin, similar to the fruit of the rock-cress genus *(Arabis)* described above, and is usually erect or spreading. Jewelflower is found below 4,000 feet in dry, open grasslands, chaparral, and open woodlands, often on serpentine, in the Coast Ranges from Monterey County northward but usually away from the immediate coast.

COULTER'S CAULANTHUS, or **COULTER'S JEWELFLOWER**, ***(Caulanthus coulteri)*** is another plant with wavy-edged, or "crisped" flower petals, and the genus shares the common name of jewelflower with the jewelflower genus *(Streptanthus)* described above. It has very similar characteristics, but the fruit can be cylindrical or flat, and the seeds are not winged as they are in *Streptanthus.* This jewelflower is an erect, hairy annual with the bases of the middle and upper leaves clasping the stem. The purple sepals can be up to an inch long but are usually shorter. They are either erect or spreading and are often a deeper purple while still in bud. The petals are longer and white to cream with purple veins and purple or brown wavy edges. The long seedpods are erect to reflexed. This jewel-flower can be found on open, dry slopes below 7,000 feet from the south-

eastern San Francisco Bay Area southward and also in the southern Sierra Nevada and the southwestern edge of the Mojave Desert.

Coulter's caulanthus, or Coulter's jewelflower

Sea rocket

SEA ROCKET ***(Cakile maritima)*** is a coastal member of the mustard family that has become naturalized from Europe. Fleshy and branched, it is prostrate to erect with deeply lobed leaves and small lavender to pink flowers with four spreading petals. The fleshy fruit has two connected segments, the upper triangular and elongate, the lower more cuplike. It is found on beach sands along the entire California coast and has largely replaced the native species, California sea rocket *(C. edentula)*, with more shallowly lobed leaves and smaller white flowers.

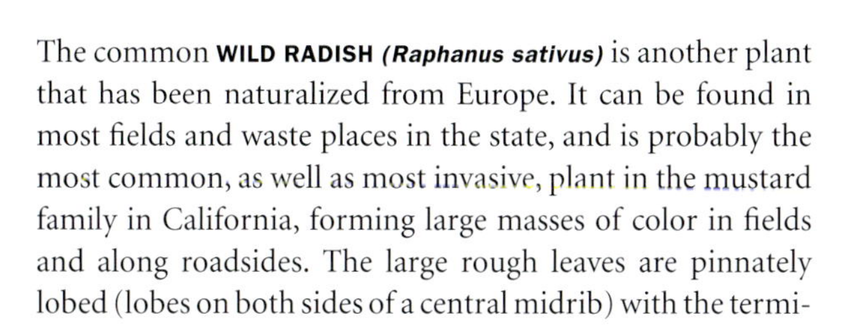

The common **WILD RADISH** ***(Raphanus sativus)*** is another plant that has been naturalized from Europe. It can be found in most fields and waste places in the state, and is probably the most common, as well as most invasive, plant in the mustard family in California, forming large masses of color in fields and along roadsides. The large rough leaves are pinnately lobed (lobes on both sides of a central midrib) with the termi-

Wild radish

nal lobe being much larger than the side lobes, much like the true mustards and other related plants. The four-petaled flowers are purple or pink with conspicuous red, brown, or purple veins. The fruit is a fat wavy pod containing several seeds. The garden radish is a cultivated form of this species, but the edible part of the wild radish is the fat curvy seedpod rather than the root. Often found growing with wild radish is jointed charlock *(R. raphanistrum),* an identical plant but with handsome creamy yellow flowers with red or brown veins and a thinner, more curvy fruit. The white form is a hybrid between these two species.

In the purslane family (Portulacaceae) a brilliant, eye-catching flower is **RED MAID *(Calandrinia ciliata),*** but the flowers generally do not open until the afternoon and not at all on cloudy days. It is a low, spreading, more or less fleshy annual with small alternate lanceolate to oblanceolate leaves and brilliant deep rose to red five-petaled flowers about one-half to one inch wide. The flowers have only two sepals, as is typical of most members of this family, and three to 15 stamens,

Red maid

and the pistil has a three-parted style. Red maid is found in sandy to loamy soil in open grasslands and disturbed areas up to 7,000 feet in most of California.

Five-stamened tamarisk

FIVE-STAMENED TAMARISK ***(Tamarix ramosissima)*** is an attractive, wispy pink shrub or small tree, but it is a pesky, invasive, water-hogging weed introduced from Europe that can dry up streams and replace native riparian vegetation if not controlled. In the tamarisk family (Tamaricaceae), it can grow up to 25 feet in height. It is a loosely branched shrub with very slender twigs, rough bark, small scaly leaves that exude salt droplets, and clusters of many, small pink or whitish flowers. The very deep roots extract water from the soil before the water can rise to a level where the more shallow-rooted native vegetation can reach it. Thus, the native plants often die off, and tamarisk can expand and take over these waterways. This species is found below 3,000 feet mostly in southern California, but it is rapidly making its way north.

INDIAN-PINK *(Silene californica),* of the pink family (Caryophyllaceae), has striking, bright red or crimson flowers about an inch in diameter. It is a perennial with a woody caudex and stems that are prostrate or decumbent (prostrate in the lower parts, but becoming erect in the upper parts). It has wide, oval leaves below but gradually smaller ones further up the stem. The flower petals are divided into four to six lobes, and each petal has two small appendages at the base. It is found in oak woodlands, chaparral, and coniferous forests through much of the state. A very

Indian-pink

similar species, but with much narrower leaves, is southern Indian-pink, or fringed Indian-pink, *(S. laciniata* subsp. *major)*, which occurs up to 7,000 feet from Santa Cruz County to Mexico. In pioneer days, Indian-pink was used as a tea for aches, sprains, and sores.

Beavertail cactus

The cactus family (Cactaceae) is well represented in the deserts of California. The plants have no leaves, but the fleshy stems are often branched into flattened, rounded segments resembling large thick leaves. Most bear long, obvious spines, but several members of the prickly pear genus, *Opuntia*, appear to be spineless. They do, however, have smaller, deciduous, barbed bristles known as glochids, as in **BEAVERTAIL CACTUS *(Opuntia basilaris)***. Although this species can have various growth forms, it is usually a low-growing plant with flat, smooth stem segments and minute rough hairs along the stem. It often has very handsome, somewhat purplish joints. Pink-magenta flowers grow along the edges of the flat branches and are stunning with their several dark magenta stamens and contrasting white or pale pink pistil. Three varieties exist, one of them *(O. basilaris* var. *basilaris)* common and occurring up to about 7,000 feet in desert and pinyon-juniper woodland habitats in much of southern California. The other two varieties are rare and very geographically limited: *O. basilaris* var. *brachyclada* has longer stem segments and occurs in the San Gabriel and San Bernardino Mountains, and *O. basilaris* var. *treleasei* bears spines and is limited mostly to Kern County.

Among the many succulent plants found in California is **SEA-FIG *(Carpobrotus chilensis)*,** in the group known as iceplant. It is in the fig-marigold family (Aizoaceae), formerly known by the more aptly descriptive name of carpetweed family, as many species in the family are mat forming and often cover large areas of beaches and sand dunes along the coast. Sea-fig is a fleshy, trailing plant with thick, erect, triangular leaves with smooth angles. The flowers, about two inches wide, are rose magenta with many thin petals whorled around a cluster of many stamens and a pistil with a large inferior ovary (attached below the sepals and petals) that forms the fig. Some debate exists about whether this species is native, but most likely it is from Africa, as is the very similar Hottentot-fig *(C. edulis)*, which has larger flowers, three to four inches wide, that are either magenta or yellow, and leaves with sharper angles. It is more common than sea-fig and is the species used to landscape highways and the like. It is also taking over many of our ocean beaches, dunes, and bluffs, crowding out many of our coastal natives. The fruit is edible but not very tasty.

Chalk dudleya

CHALK DUDLEYA ***(Dudleya pulverulenta)*** belongs to a large group of succulents in the stonecrop family (Crassulaceae). The genus *Dudleya* has the common name liveforever and is sometimes called hen-and-chickens because of the habit of some species to proliferate by sending out new young plants from the base. Chalk dudleya is one of the showiest and largest of our native liveforevers, growing to more than a foot high with broad leaves covered with a dense white mealy powder. The petals are deep red and half-an-inch long or more. It is a coastal species found up to 5,000 feet on rocky cliffs and in canyons from San Luis Obispo County southward, flowering from May to July.

BEACH SAND-VERBENA ***(Abronia umbellata)*** is a trailing plant in the four-o'clock family (Nyctaginaceae), a family with opposite leaves and bell-shaped or trumpet-shaped flowers with petal-like sepals but no actual petals. This species is one of the more common ones, being a perennial with stems up to three feet long, and having mostly rose-colored flowers in dense

Beach sand-verbena

clusters enclosed by several bracts. It is found on coastal strand from Del Norte County to San Diego. With it may grow a stouter, glandular plant with smaller darker flowers, red sand-verbena *(A. maritima).* Another sand-verbena species is described in "Yellowish Flowers."

In the buckwheat family (Polygonaceae) is a genus of low-growing, often pink-flowered plants called spineflowers. **TURKISH RUGGING *(Chorizanthe staticoides),*** an erect annual in this group, is usually about five to eight inches high but sometimes taller and has oblong basal leaves densely hairy on the undersides. The numerous small flowers are in dense clusters on terminally branched stems. Each flower sits in an involucre, or tubelike structure, that spreads into six tapering lobes above, each tipped with a hooked spine. The flower has no petals but is composed of five or six very petal-

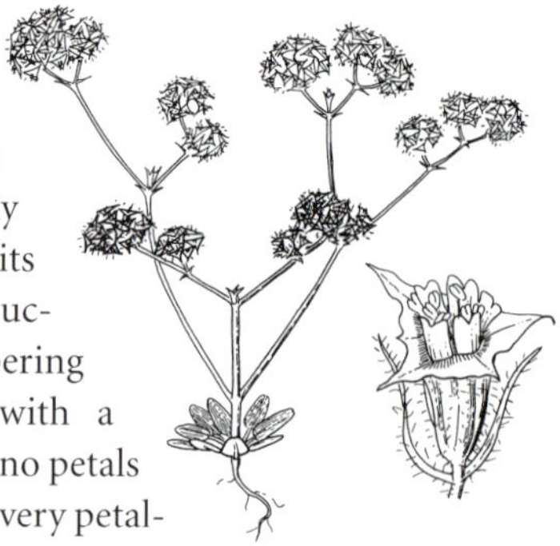

Turkish rugging

Pink spineflower

like, unequal sepals ranging from pink to white and having entire edges, not fringed or toothed as in some similar species. Turkish rugging occurs below 6,500 feet in coastal scrub, chaparral, and desert scrub from central California southward.

PINK SPINEFLOWER *(Chorizanthe membranacea)* is related to Turkish rugging *(C. staticoides)* but has longer stems and woollier flowers. It also differs from the above species in having a mostly translucent, membranous, somewhat hairy involucre. The flower inside the involucre is usually densely hairy. The plant is often grayish with a few basal leaves and alternate leaves along the stem. It is more sparsely flowered than the above species, having only a few flowers in the lower parts but more above. It is common in dry grasslands, chaparral, foothill woodland, and other dry habitats below 5,000 feet in much of California.

In the same family is a very large genus, wild buckwheat *(Eriogonum). Eriogonum* is Greek for "woolly knees," referring to the hairy nodes on the stems of some plants. More than 100 native species of wild buckwheat are found in California, and they differ from the spineflowers above in having cup-shaped or cylindrical involucres that enclose clusters of flowers rather than a single flower. Many species are perennial, but **SLENDER ERIOGONUM *(Eriogonum gracillimum)*** is a slender annual that begins to bloom in middle to late spring and often continues through the summer months. The leaves are mostly basal, but there may be a few on the stem as well. The bell-shaped involucres are on small thin stalks, and the five petal-less flowers are composed of rose to white, wavy-edged petal-like sepals. It is found in gravel to clay habitats between 1,000 and 4,000 feet in the southern Coast Ranges, the southern Sierra Nevada, and the Transverse Ranges. Several other annual species of wild buckwheat can be found around the state, sometimes in very large numbers, blooming mostly in summer and early fall.

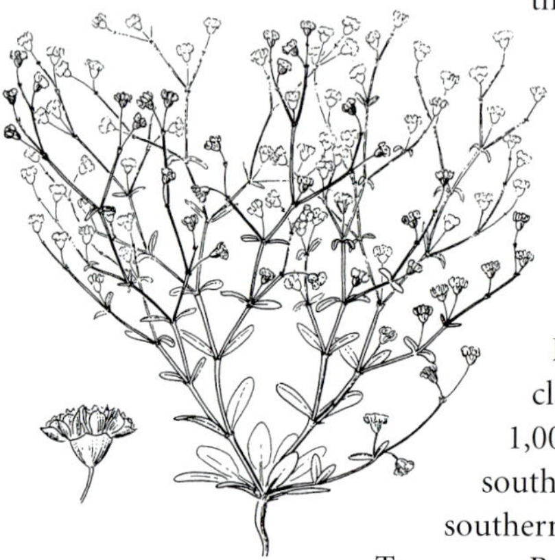

A very different looking group of plants in this family is dock, related to the garden rhubarb *(Rheum).* Not really wildflowers per se, but very unique in flower and fruit structures, these plants occur throughout the state and often arouse a hiker's curiosity. **CURLY DOCK *(Rumex crispus)*** is a nonnative perennial from Europe with a tall stout central stem with several erect branches. The widely-lanceolate leaves are undulate or wavy, appearing crisped or curled. They are clustered at the base of

Curly dock

the stem, as well as scattered along it, and the thick leaf stems are deeply grooved on the upper side as in rhubarb. The numerous small green flowers with multiple stamens and pistils soon produce strange three-parted fruits with small, seedlike growths called tubercles in the center of each segment. The size and quantity of these tubercles are important in differentiating between the several species of dock. Curly dock has three oval tubercles, one on each segment and less than one-third the width of the segment. The fruit turns a deep rusty brown in summer, and its tall stalks are conspicuous in seasonally wet fields and disturbed areas throughout the state up to about 8,000 feet. The leaf is edible, but large quantities can be toxic because it is very high in vitamin A. When crushed between the fingers and applied to the skin, the leaf can also be used to relieve the itch of an insect bite or a stinging nettle irritation.

Long-tailed wild-ginger

Plants in the pipevine family (Aristolochiaceae) have petal-less flowers with three sepals and are often aromatic. **LONG-TAILED WILD-GINGER *(Asarum caudatum)*** is a perennial herb with a creeping stem and large heart-shaped dark green leaves that give off a spicy odor when crushed. The dark maroon flowers are at ground level, hidden below the leaves. The three sepals are fused into a short, wide, bowl-like tube below and spread into wide, rounded lobes that taper into very long, tail-like tips. It is found in moist, shaded areas within forests below 7,000 feet from the San Francisco Bay Area northward to British Columbia.

Dutchman's pipe

DUTCHMAN'S PIPE ***(Aristolochia californica)*** is a vine in the same family with an unusual tubular, curving flower resembling the bowl of an old-fashioned pipe. It is a woody climbing plant eight to 12 feet high and usually found in low ground growing over bushes. The more or less heart-shaped leaves are two to six inches long. Although some of the tropical species in this genus are brilliantly colored, this species is more muted with brownish or green flowers lined with pink or red. Although it generally grows near streams in forests and woodlands, it is occasionally found in open grasslands. It can be found up to 2,000 feet, ranging from Monterey County northward in the Coast Ranges, and in the Sierran foothills from Sacramento and El Dorado Counties to the head of the Sacramento Valley. It begins to flower in January.

FRINGE CUPS ***(Tellima grandiflora),*** in the saxifrage family (Saxifragaceae), is aptly named because the flower forms a cup and the spreading petal lobes are laciniately divided into many narrow segments, thus appearing very fringed. It has

Fringe cups

both basal and stem leaves, although the latter get progressively smaller further up the stem. The many flowers are situated along the upper 16 to 40 inches of the stem and can be green, white, or pink, all on the same flowering stalk. It is found in moist woods, along streams, and in rocky places below 7,000 feet from central California northward to Alaska.

The gooseberry family (Grossulariaceae) consists of only one genus, *Ribes,* which is composed of gooseberries with spines and currants without spines. **CHAPARRAL CURRANT *(Ribes malvaceum)*** is in the latter group and has hanging clusters of 10 to 25 pink flowers. The rounded leaves are one to two inches long and have a lovely, musky fragrance. The flowers have long pink sepals, smaller erect pale pink or white petals, and inferior ovaries that become the dark blue to black, bitter-tasting currants. It occurs in chaparral and dry oak wood-

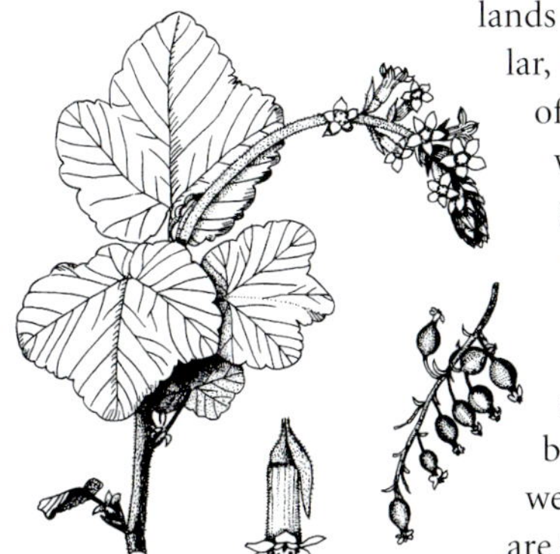

lands below 5,000 feet. Very similar, but occurring in moister, often more shaded habitats with cooler, generally foggier climates, is red flowering currant *(R. sanguineum)*, differing mostly in having redder sepals and a style that is not hairy at the base. It is widely cultivated in western Europe. Both species are found in central and northern California, but chaparral currant extends into southern California and the Channel Islands as well. Both are early bloomers, but chaparral currant can be found in flower as early as October or November, whereas red flowering currant usually begins to bloom between January and March.

Two common species in the gooseberry group are **HILLSIDE GOOSEBERRY *(Ribes californicum)*** and **CANYON GOOSEBERRY**

Hillside gooseberry

(R. menziesii). They differ from the currants in having spines on the stem; thinner, nonaromatic leaves; only one to three flowers in a cluster; larger flowers with reflexed sepals and short white petals; and glandular, bristly berries. In hillside gooseberry, the underside of the leaf is smooth, and three long spines occur at the leaf nodes where the leaf stem attaches to the main stem. The flowers have green to red sepals, and the berries have bristles of unequal lengths. In canyon gooseberry the leaves are glandular on the underside, and there are small spines all along the stem as well as the three long ones at the leaf nodes. The flowers have deep purple sepals, and the bristles on the berries are all of equal lengths. Both species occur in woodlands, forests, and chaparral below 3,000 feet, mostly in the Coast Ranges, but canyon gooseberry is also found in the southern Sierra Nevada. A less common variety of hillside gooseberry, *R. californicum* var. *hesperium,* has hairy leaves and occurs in southern California in the western Transverse Ranges and the San Gabriel and Santa Ana Mountains.

Fuchsia-flowered gooseberry

One of the most beautiful native gooseberries, and one that is often found in cultivation, is **FUCHSIA-FLOWERED GOOSEBERRY**

(Ribes speciosum). It is a spiny shrub three to six feet tall with long, spreading, bristly branches, shining green leaves, and clusters of one to four bright red hanging flowers with four, instead of five, sepals and petals. The berry is densely glandular and bristly. It is common in shaded canyons below 1,500 feet near the coast from Santa Clara County southward. It is a striking plant and another early bloomer, beginning to flower shortly after New Year's Day.

Common Pacific pea

In the pea family (Fabaceae), a very close relative of the popular garden sweet pea is **COMMON PACIFIC PEA** ***(Lathyrus vestitus),*** with the same long, weak, climbing stem, and compound leaves with the midrib ending in a curling tendril. The stem has much narrower, if any, wings along its edges than sweet pea, and the compound leaves have several more leaflets, usually from eight to 12. A long flower stem arises from the base of the leaf stem with eight to 15 crowded pale lavender to dark purple flowers with the typical two-lipped pea family flower consisting of a large upper banner petal, two side wing petals, and a keel petal hidden between the two wings. It occurs below 5,000 feet in various habitats, usually shaded, in much of California. It has three varieties varying mostly in flower color and habitat. Other species of wild pea *(Lathyrus)* have white to yellowish or red flowers, but none have the lovely fragrance of the cultivated sweet pea.

ROUND-LEAVED PSORALEA *(Hoita orbicularis)* is a creeping perennial herb in the pea family. Growing from a prostrate stem, the long-stemmed leaves are divided into three rounded, gland-dotted leaflets about one to four inches long that give off a heavy scent. The flower stalks rise directly from the creeping stem and are one to two feet tall with several brown to reddish purple pea-shaped flowers, each about half-an-inch long. It can be found below 7,000 feet in moist meadows and hillsides and along streams throughout California. Another common species found in the same habitat is leather root, or California hemp, *(H. macrostachya)*, which is much taller and has erect stems.

Round-leaved psoralea

Chaparral-pea

CHAPARRAL-PEA *(Pickeringia montana)* is a beautiful magenta-flowered shrub, usually about six feet tall. It has a complex branching pattern, thick sharp thorns, and small three-parted leaves. The showy flowers are almost an inch long and are either at the tips of the branches or in the leaf axils. The fruits are wavy, oblong pods about one to three inches long with one to eight seeds inside. The plant can range from glabrous (no hairs) to densely hairy. It is usually found below 5,500 feet in chaparral on dry slopes and ridges but can also occur in dry woodlands and washes. It ranges from Mendocino and Butte Counties to San Diego.

Mallow, hollyhock, cotton, and hibiscus are familiar names to a plant lover. They are all members of the mallow family (Malvaceae), one that is well represented in California. In this family, the leaves are usually very wide and rounded. The five-petaled flowers often have a set of bracts (leaflike structures) below the sepals, and the multiple stamens are fused into a tube around the pistil. **WHITE-COAT MALLOW *(Malacothamnus fremontii)*** is a woody shrub in this family, covered with white woolly hairs, giving it a grayish appearance. It has coarse, stiff

White-coat mallow

branches and oval leaves about two inches long. The pink to light purple flowers are almost an inch across with five narrow bracts below the sepals. It is found in chaparral and pine woodlands below 7,500 feet through most of California, often more abundantly after a fire.

CHECKER MALLOW ***(Sidalcea malvaeflora)*** is a highly variable plant in the mallow family with lovely pink flowers. It is an herbaceous perennial growing from a rhizome and often has a woody base. The hairy stem can be from six inches to two feet long, and the leaves are quite variable. In general the basal ones are slightly lobed or toothed, and the upper are smaller and more deeply lobed. The flowers range from pale to deep pink with five filmy petals about half-an-inch to almost an inch long, strongly veined and squared off at the top. It occurs up to 7,500 feet in grasslands, woodlands, and scrub throughout most of California. Eight subspecies, varying mostly in leaf shape and size, types of hairs, and fruit composition, occur. More delicate, and an annual, is fringed sidalcea *(S. diploscypha),* with larger, darker pink flowers clustered near the top of the stem and slightly fringed petal tips.

Checker mallow

A third member of the mallow family is **PARRY'S MALLOW *(Eremalche parryi)***. An erect annual that can grow up to 20 inches tall, it usually has stems with star-shaped hairs, especially near the tips. The leaf is one to two inches wide and deeply divided

Parry's mallow

into three to five lobes. The large five-petaled flowers range from white to purple, and each petal can be up to an inch long. It is found in grassland, scrub, and woodland below 4,500 feet in the central and southern Sierra Nevada and the Tehachapi Mountains and from the eastern San Francisco Bay Area and San Joaquin Valley southward through the inner Coast Ranges and Central Valley to the western Transverse Ranges.

REDWOOD-SORREL *(Oxalis oregana),* in the oxalis family (Oxalidaceae), is a delicate perennial growing from a fleshy, creeping rhizome. The stem, if any, is very short, but it has a cluster of several leaves with long hairy stems up to eight inches long. The thin, delicate leaves are divided into rounded, cloverlike lobes that fold downward at night. The five-petaled flowers are also delicate, range in color from white to deep pink, and have stems up to six inches long. The spreading petals are up to an inch long, and the 10 stamens are of two lengths, five short and five long. It is found in moist, shaded redwood and other coniferous forests below 3,500 feet from Monterey County northward to Washington.

Redwood-sorrel

California geranium

The geranium family (Geraniaceae) is well represented in California and easily recognized by the long, pointed, beaklike fruit that separates into five narrow segments when mature, each with a seed attached at the bottom. **CALIFORNIA GERANIUM** ***(Geranium californicum)*** is a perennial with palmately lobed leaves and soft, hairy light pink flowers. The five petals are each about an inch long, veined with purple, and often notched at the center of their wide tips. This species occurs near streams and in moist meadows and woodlands from 3,000 to 9,000 feet in the Transverse and Peninsular Ranges of southern California and in the Sierra Nevada. In northern California, Oregon geranium *(G. oreganum)* with larger, darker pink, glabrous flowers on shorter stems, can be found in similar habitats.

Unfortunately, several small widespread and very invasive nonnative weeds in California are also in the geranium family. **CUT-LEAVED GERANIUM** ***(Geranium dissectum)*** has small flowers with five red pink petals that overlap at the base. The sepals have a short bristle at the tip that helps differentiate this species from the dove's-foot geranium *(G. molle)* described below. The lobes of the deeply palmately divided leaves are toothed or sometimes lobed themselves. The beaked fruit of most species in this genus is about an inch long, giving the genus its common name of cranesbill. It is found in open or shaded disturbed sites up to 4,000 feet throughout the state.

Cut-leaved geranium

DOVE'S-FOOT GERANIUM ***(Geranium molle)*** is another small widespread weed in the geranium family. The leaf is palmately lobed, as are all plants in this genus, helping to distinguish them from those in the filaree genus, *Erodium*, described below. In this species, the leaves are only shallowly lobed and are generally very soft hairy. The flower is usually smaller and a rosier pink than the cut-leaved geranium *(G. dissectum)* described above.

Dove's-foot geranium

It is slightly cupped and the sepals are pointed but lack the small bristle seen in the above species. It too is found in disturbed sites but also in less-disturbed grasslands below 3,500 feet throughout California.

Also in the geranium family is **RED-STEMMED STORKSBILL**, or **RED-STEMMED FILAREE**, ***(Erodium cicutarium)***, another very widespread and invasive weed. It is usually only a few inches tall, and its leaves are deeply dissected into several side leaflets that may also be dissected, giving the leaf a feathery or fern-like appearance. The five oval, spreading pink petals are about a quarter-inch long each, with short white hairs protruding from both sides of the base of the petal. The pointed fruit beak is about two inches long. Musk filaree, or white-stemmed filaree, *(E. moschatum)* is similar but can be up to two feet tall. The leaves are divided into toothed, not dissected, leaflets, and there are no hairs at the base of the petals. Both species are found below 6,500 feet in open, mostly disturbed sites throughout California.

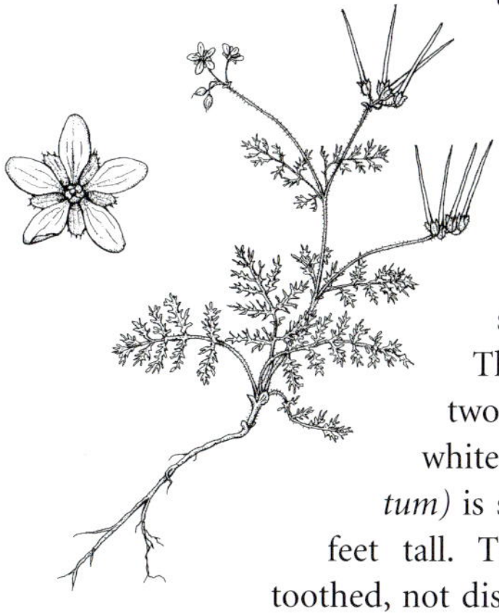

LONG-BEAKED STORKSBILL, or **LONG-BEAKED FILAREE**, ***(Erodium botrys)*** is another very common member of this genus. It has thicker, darker green leaves than the above species that may be toothed to deeply lobed but usually not divided into separate leaflets. The flower is cuplike, larger, and generally a deeper pink. The fruit has a very long beak, up to eight inches long. It is found in disturbed sites below 3,500 feet throughout California. The profusion of this and the above species can be accounted for by their interesting seed dispersal method. As

Long-beaked storksbill, or long-beaked filaree

each segment of the mature beak pulls apart, it coils into a twisted tail with a seed at the end. This coil often gets caught in the hair of passing animals or the clothes of hikers and is transported to other areas. When it finally falls to the ground, the tail coils and uncoils like a corkscrew in response to alternating wet and dry conditions, thus planting the seed further and further into the ground with each twist.

In the evening-primrose family (Onagraceae) is a taller, native plant, **DENSE-FLOWERED BOISDUVALIA** ***(Epilobium densiflorum),*** with erect, sometimes branched, stems, usually between two and four feet tall. The lanceolate, hairy leaves are alternate and crowded on the stem. The sometimes glandular inflorescence of many rosy pink flowers is interspersed with leaves, the flowers sometimes being almost hidden by the leaves, but more commonly forming an attractive long pink spike of many small, crowded flowers with four notched petals and long inferior ovaries. It appears in late spring near streams and washes and in seasonally wet grassy fields below 8,500 feet through much of California.

Dense-flowered boisduvalia

Northern willow-herb

Also found in wet places is **NORTHERN WILLOW-HERB *(Epilobium ciliatum)*,** with wider, lanceolate, opposite leaves, and flowers clustered at the top of the plant rather than in the long, thick spikes of the above species. The individual flowers are similar to the above-described dense-flowered boisduvalia *(E. densiflorum)*—pale pink to rosy purple, less than an inch long, four petals notched at the tip—but the petal lobes are generally more spreading and the inferior ovaries are longer and narrower. The seeds inside the fruit have tufts of white hairs attached at their tips, as do the seeds in the willow family, hence the name willow-herb. It occurs near streams, in moist meadows, and in disturbed wet areas up to 13,500 feet throughout California and most of North America.

In the same family is the clarkia genus *(Clarkia)*, named for Captain William Clark of the Lewis and Clark Expedition to

Elegant clarkia

the Pacific Northwest early in the nineteenth century. **ELEGANT CLARKIA** ***(Clarkia unguiculata)*** can be more than three feet tall and very widely branched. The four spreading petals are narrow below and more or less triangular or diamond shaped above. They are half-an-inch to an inch long, and the color can range from pale pink to salmon to almost burgundy. The eight stamens have red anthers and surround a pistil with a four-parted, crosslike stigma that extends well beyond the stamens. The ovary below has eight grooves and is sometimes hairy. It is found in woodlands below 5,000 feet in much of California, beginning to bloom in May.

Another clarkia that has wider, entire petals is **LARGE GODETIA** ***(Clarkia purpurea*** **subsp.** ***viminea).*** The leaves are linear to narrowly lanceolate, and the flowers, ranging from pink to burgundy, have petals from half-an-inch to an inch long, often with a red or purple spot just above the middle. The four-

Large godetia

parted stigma at the top of the pistil extends beyond the eight stamens, and the long thin ovary has eight grooves. The sepals are fused into groups of two, spreading to either side of the stem beneath the petals. It is found below 5,000 feet in grassy and brushy or wooded places in most of California but is more common in the Sierra Nevada and the more southern mountains.

Very closely related to the above species is **FOUR-SPOT *(Clarkia purpurea* subsp. *quadrivulnera)*.** It is very similar to large godetia *(Clarkia purpurea* subsp. *viminea)*, but the flowers and the plant itself are often smaller. The flowers are almost never crowded or clustered as they are in the larger subspecies, and they are generally a paler pink or almost lavender. In spite of the common name, the petals often lack the central spot. The stigma does not extend beyond the stamens in this subspecies, but the ovary is still eight grooved, and the sepals are usually fused into groups of two, although they can sometimes all be separate. It is found in open grassy and brushy or wooded places up to 5,000 feet in much of California, as is large godetia, but it is more common at lower elevations. Very similar, but blooming later, is small clarkia *(C. affinis)*, which sometimes has several small purple dots on the petals and has all four sepals fused together and spreading to one side of the stem.

Four-spot

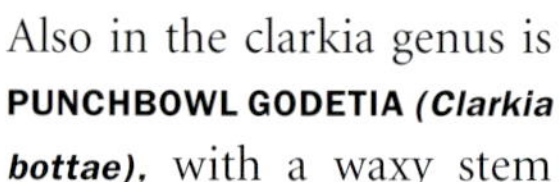

Also in the clarkia genus is **PUNCHBOWL GODETIA *(Clarkia bottae)*,** with a waxy stem and lanceolate leaves. The flower is more bowl shaped than spreading, and the four sepals are fused together and leaning to one side, as in small clarkia *(C. affinis)*. Before the flowers open, the buds are reflexed, which helps differentiate it from the above species, which all have erect buds. It is found in openings in chaparral, woodland, and coastal scrub below 3,000 feet in much of southern California except in Monterey County, where a very similar but rare species occurs, Jolon clarkia *(C. jolonensis)*, which is usually taller and has smaller flowers and gray rather than brown seeds. A somewhat similar species common in the northern part of the state, but with erect buds and a basal red spot on each petal, is farewell-to-spring *(C. amoena)*. Many species of clarkia have been widely cultivated, especially in northern Europe, where double flowered and other forms have been developed.

Punchbowl godetia

CALIFORNIA MILKWORT ***(Polygala californica)*** looks like it might be in the pea family at first glance, but a closer look reveals that it is actually a member of the milkwort family (Polygalaceae), which has flowers with five sepals, the two large side ones being colorful and very petal-like. It has only three actual petals—a lower one shaped like a keel but with a small blunt tip, or beak, and two upper, usually narrower ones. In this species, the flowers are rose colored and about half-an-inch long. The plant is about a foot high or less and has several thin oval leaves about half-an-inch to almost three inches long. It is found in brush and open woods on rocky ridges and exposed slopes below 4,500 feet from San Luis Obispo County northward and also in the northern Channel Islands. A closely related and similar species, Sierra milkwort *(P. cornuta)*, ranges the length of the state, in spite of its common name, but in more scattered populations. The beak on the keel petal is more linear in this species, and it is not notched or contorted as it is in California milkwort. At one time, plants in this family were thought to be helpful in increasing the flow of milk.

California milkwort

SUGAR BUSH *(Rhus ovata)* is an evergreen shrub related to sumac *(R. glabra)* and poison-oak *(Toxicodendron diversilobum)* in the sumac or cashew family (Anacardiaceae). About six feet tall or more, it has stout reddish twigs and thick leathery leaves two to three

Sugar bush

inches long with smooth edges. During winter the branches have attractive clusters of numerous small reddish flower buds that, in spring, become a myriad of tiny pink or white five-petaled flowers with red sepals. The fruit is flat, glandular, and reddish. It grows on dry slopes from Santa Barbara County to Baja California and is most abundant in the hot interior valleys. It is also found on Santa Cruz and Catalina Islands. A close relative is lemonadeberry *(R. integrifolia)*, which can be distinguished by its flat, oblong leaves, whereas the leaves of sugar bush are ovate and somewhat folded upward along the midrib. It grows in chaparral areas, usually on north-facing slopes and in canyons below 3,000 feet in southwestern California and into Baja California. Both species have an acid substance on the surface of the flattened fruits.

Sweet-shrub, or spicebush

Another notable shrub is **SWEET-SHRUB**, or **SPICEBUSH**, ***(Calycanthus occidentalis)***, in the sweet-shrub or calycanthus family (Calycanthaceae). The only native member of this family in California, it usually has a rounded shape and grows from three to 15 feet tall. The large deciduous leaves, up to six inches long, are opposite each other on the stem and have a very pleasant aroma when bruised. The large, solitary flowers are a deep reddish brown and measure about two inches across, consisting of numerous spirally arranged perianth parts (petals and sepals that are identical and indistinguishable from each other). The woody oval fruit is almost two inches long and has an attractive veining pattern on the sides. Found below 5,000 feet in moist, shady places in canyons and near ponds and streams, it occurs in the Cascade Range and northern Coast Ranges and along the west base of the Sierra Nevada from Tulare to Shasta Counties. It flowers from April to August.

The carrot family (Apiaceae) is another large family in California. The flowers are always in umbels (arising from the same point, usually the tip of the stem) and can be open, with each individual flower having a stem, or closed, with each flower attached at its base directly to the central point, as in **PURPLE SANICLE**, or **SHOE BUTTONS**, ***(Sanicula bipinnatifida)***. Most members of the family have white or yellow flowers, but this species is an exception. The densely flowered round heads are a beautiful deep red purple that is very striking in the grassy green fields where it is usually found. Each head consists of about 20 individual flowers a quarter-inch wide or less with five long, dark stamens and a two-styled pistil. Several flowers in each head usually have very small stems (two millimeters or less), but most are stemless. The entire plant is quite hand-

Purple sanicle, or shoe buttons

some, with deeply dissected, sometimes purplish leaves divided almost to the midrib into several coarsely and unevenly toothed segments. It is found in open grasslands and pine woodlands below 4,000 feet through most of California.

Cycladenia

Although fairly uncommon, **CYCLADENIA** ***(Cycladenia humilis)*** is a very appealing perennial herb in the dogbane family (Apocynaceae). It is a fleshy and glaucous plant with a milky sap and two to five pairs of large oval to round leaves that can be more than three inches long. The glabrous, rose purple flowers are tubular with spreading, rounded, wavy petal lobes. The seedpods are more than two inches long, and the seeds inside have long tufts of hairs. It usually grows in patches in rocky or sandy places between 4,000 and 9,000 feet in the Klamath Mountains, the North Coast Ranges, and the northern Sierra Nevada. In the southern part of the state is found a different variety, *C. h.* var. *venusta,* in which the flowers are soft hairy, especially on the outside. It grows in the San Gabriel Mountains, the Santa Lucia Range, and the east side of the Sierra Nevada.

California milkweed, or round-hooded milkweed

Another family of plants with milky sap is the milkweed family (Asclepiadaceae) with unique flower structures. The five sepals and five petals are generally reflexed, and the stamens are fused to the style of the pistil, forming a central column. The five anthers are on the outside of the column, near the top, and between each of them is an expanded structure. **CALIFORNIA MILKWEED**, or **ROUND-HOODED MILKWEED**, ***(Asclepias californica)*** is a gray, very woolly-haired perennial in this family, growing up to two feet high with large, opposite leaves. The flowers are pink to purple or maroon, and the expanded structure between the anthers on the central column is hoodlike, dark purple, and situated lower than the anthers. The fruit is narrow and podlike with numerous silky-haired seeds. California milkweed is found in grassy flats and brushy hillsides below 7,000 feet through much of California west of the Sierra Nevada and southward to Baja California. Some of the milkweeds were used in early days for healing cuts and treating rheumatism.

THRIFT, or **SEA-PINK**, ***(Armeria maritima* subsp. *californica)*** is a tufted plant in the leadwort family (Plumbaginaceae) with many linear, flat leaves at the base. The dense head of rose pink flowers at the top of a foot-tall flower stalk looks very much like an onion flower head, but a close look reveals that the flower parts are in groups of five, not the groups of three or six characteristic of onion flowers. The individual flowers are funnel shaped and about one-third inch long, and the sepals and petals are both pink. It is found in coastal sandy places and on sea bluffs from San Luis Obispo County northward.

Thrift, or sea-pink

California rose-bay

CALIFORNIA ROSE-BAY ***(Rhododendron macrophyllum),*** in the heath family (Ericaceae), is an evergreen shrub three to 10 or more feet tall with coarse twigs and leathery dark green leaves. The flowers are more than an inch long, broadly bell shaped, and rose to rose purple or, on rare occasions, white. Growing in dryish to damp, more or less shaded coniferous forests below 4,000 feet, it is found near the coast from Monterey to Del Norte and Siskiyou Counties. It blooms from April to July, adding much color to the forests of the Redwood region.

PACIFIC STARFLOWER ***(Trientalis latifolia)*** is a small delicate perennial in the primrose family (Primulaceae). Plants in this family have simple leaves, and the flowers are four- or five-parted with spreading or reflexed petals and deeply lobed calyces (fused sepals). Pacific starflower is a diminutive and dainty member of the family with a whorl of ovate leaves at the top of a short stem and two or three small pink flowers rising from the center of the whorl on threadlike, almost invisible stems. This ethereal flower dots forest and woodland floors below 4,500 feet in spring in the Coast Ranges, the Sierra Nevada, and the Mount Shasta area in northern California.

Pacific starflower

Padres' shooting star

Also in the primrose family, but with reflexed rather than spreading petals, is **PADRES' SHOOTING STAR *(Dodecatheon clevelandii)***. Related to cyclamen (*Cyclamen* spp.), it has several oblanceolate leaves at the base of a leafless flowering stalk. The one to 16 magenta to white flowers at the top of the stalk have five reflexed petals up to an inch long with bands of yellow, white, or both at the base, which form a colorful ring at the center of the flower. Five broad, dark stamens, usually with yellow markings, protrude downward from the center. Padre's shooting star grows on grassy slopes and flats below 1,500 feet, sometimes forming large, impressive purple carpets. It can be found in the Sierra Nevada and from central and southern California to Baja California. Four subspecies occur, varying mostly in the placement of the yellow markings on the stamens. Several other species of shooting star occur in California, varying in petal and stamen coloring.

A small delicate pink annual in the gentian family (Gentianaceae) is **CANCHALAGUA**, or **CENTAURY**, ***(Centaurium venustum)***. Usually less than a foot tall, it has small, opposite leaves and several pink flowers in either open or dense clusters. The flowers have narrow tubes that spread out into five petal lobes about half-an-inch long. The stamens have distinctive twisted yellow anthers. It occurs in dry scrub, grassland, and forest below 4,500 feet in much of the state. About six other species occur in the state, mostly in wetter areas, and several of them intergrade and can be difficult to distinguish from one another.

Canchalagua, or centaury

One of the more conspicuous families in California is the phlox family (Polemoniaceae), and among its spring bloomers is **GREAT POLEMONIUM** ***(Polemonium carneum)***. A hairy perennial one to two feet high, it has compound leaves

Great polemonium

divided into three to 10 pairs of lanceolate to ovate leaflets. The three to seven purple to pinkish flowers are in an open cluster with each flower being about one-half to one inch in diameter and consisting of a short bell-shaped tube spreading into five large petal lobes above. The five stamens are hairy at the base, and the long pistil extends far beyond the stamens. It is found below 6,000 feet in grassy and brushy places from central California northward to Washington.

SHOWY PHLOX *(Phlox speciosa* subsp. *occidentalis)* is an erect, branched perennial with long, thin opposite leaves. The bright pink flowers are up to an inch in diameter, with narrow tubes about half-an-inch long and petal lobes ranging from notched to deeply lobed. The stamens are very short and at-

Showy phlox

tached at two different levels in the flower tube (which can best be seen by holding the flower up to the light), and the pistil has a stigma longer than the style. It grows on wooded slopes and rocky places from 1,500 to 7,500 feet from Sonoma County northward and along the base of the Sierra Nevada from Fresno County northward.

The only annual plant in the phlox genus, *Phlox,* is **SLENDER PHLOX *(Phlox gracilis)*.** Only a few inches tall and often inconspicuous, it can have only a single stem or be very branched. It is generally glandular hairy, at least in the upper parts. The leaves are narrow and mostly opposite, but they can often be alternate on the upper sections of the stem. The flower tube is half-an-inch long or less, and the rosy pink petal lobes are only one to two millimeters long. It is found in moist or dry places in many different habitats up to 11,000 feet in most of California.

PRICKLY-PHLOX *(Leptodactylon californicum)* is a low, somewhat spiny shrub related to plants in the true phlox genus but easily distinguished by the stamens all being attached at the same

Slender phlox

Prickly-phlox

level in the flower tube, whereas in the phlox genus they are at two different levels, as noted above. Prickly-phlox is one to three feet tall, somewhat spreading, and may be woolly or glandular. The leaves are in crowded clusters and divided into linear, spiny lobes. The narrow flower tube spreads into large pink petal lobes above. It is found on dry slopes and banks below 5,000 feet from San Luis Obispo County to the Santa Ana and San Bernardino Mountains. It flowers from March to June.

A final pink-flowered member of the phlox family is **GROUND-PINK**, or **FRINGED-PINK**, ***(Linanthus dianthiflorus),*** a small annual with very thin, linear, opposite leaves on a thin, hairy stem two to five inches tall. The funnel-shaped flower tube is pink below and yellow above, and the petal lobes, about half-an-inch long, are pink or white with purple markings and toothed at the tips, giving them a fringed or pinked appearance. It is common in open sandy places below 4,000 feet from Santa Barbara to San Diego, beginning to bloom shortly after Christmas.

Ground-pink, or fringed-pink

Rigid hedge-nettle

RIGID HEDGE-NETTLE ***(Stachys ajugoides* var. *rigida)*** is in the mint family (Lamiaceae), in spite of its common name, which it probably received because some plants have very stiff, sharp hairs similar to those found on nettle plants. It is a perennial herb with a square stem and opposite, oval to lanceolate, sometimes hairy leaves, generally with rounded rather than pointed teeth along the edges. The flowers are pink or white with pink or purple markings and are whorled around the upper stem in tiers, usually with several small leaves between them. The tubular base of the flower opens up into two lips, the upper two-lobed and the lower three-lobed with the middle lobe much longer than the two side ones. It is common in many habitats throughout California below 8,000 feet. Closely related is bugle hedge-nettle *(S. ajugoides* var. *ajugoides),* with thinner, gray-hairy, often glandular leaves with wedge-shaped bases, and fewer, paler flowers. It is found in moister, open habitats through much of the state. A third common member of the hedge-nettle genus, *Stachys,* that looks similar and occurs in similar habitats is California hedge-nettle *(S. bullata),* with larger, more deeply colored flowers and stiff reflexed hairs on the angles of the stem. It is usually near the coast and occurs from central California southward and in the Channel Islands.

Mustang mint

MUSTANG MINT *(Monardella lanceolata)* is an erect annual one to two feet tall, with a pleasant odor to the crushed foliage. It is usually branched in the upper parts, and the stem is often purplish and glandular or hairy with lanceolate leaves one to two inches long. The flowers are in several rounded heads about an inch wide or less. Below each head are several, green, more or less papery bracts with purple tips and cross veining. The individual flowers are about half-an-inch long, rose purple or paler, and have long stamens extending out of the flower. It is found in dry, open, rocky places in chaparral and woodland below 8,500 feet in the Sierra Nevada and from central California southward to Baja California.

MOUNTAIN-PENNYROYAL ***(Monardella macrantha)*** is a creeping, mat-forming plant with red tubular flowers more than an inch long. Its shiny green leaves are leathery and an inch long or less. The bracts below the flower head are oblong to elliptic, the outer ones being very leaflike, and the inner bracts hairy, papery thin, and often red. It grows from 2,500 to 6,000 feet in the mountains from Monterey County to Baja California and blooms from April to July. The crushed foliage of most species of *Monardella* has a fresh, clean, minty odor.

Mountain-pennyroyal

PITCHER SAGE ***(Salvia spathacea)*** is a very handsome plant with clusters of large brilliant red to salmon flowers. It is a coarse perennial herb with underground rootstocks and large deeply veined leaves up to eight inches long, sparsely hairy above and densely hairy on the underside. The flower tube is an inch long or more and opens into a shallowly two-lobed upper lip and a half-an-inch long, three-lobed lower lip, with long stamens extending far beyond the flower tube. The purplish bracts below the flowers are often quite conspicuous and

Pitcher sage

showy. It is found in grassy and shaded places in many habitats below 2,500 feet and ranges from Sonoma to Orange Counties.

Closely related to the mint family is the figwort family (Scrophulariaceae), also characterized by a tubular two-lipped flower, but the three lobes of the lower lip are more or less equal in length. The stem is usually round, not square, and the leaves can be opposite or alternate on the stem. The large genus *Castilleja*, named for a Spanish botanist, consists of two groups of plants, Indian paintbrush and owl's-clover. The first has mostly reddish flowers, and the second has mostly pink or yellow flowers, but both have spikelike inflorescences of several tubular flowers with long hooded or beaked upper lips and very small lower lips. Both are parasitic on the roots

Lay-and-Collie's Indian paintbrush

of other plants. **LAY-AND-COLLIE'S INDIAN PAINTBRUSH *(Castilleja affinis)*** is a branched perennial six to 24 inches tall with narrow leaves usually roundly lobed at the tips. Each flower has a red or yellow bract below it that may be entire or lobed. The red calyx tube is divided halfway down in the front and back but only about one-third on each side. The flower tube is about an inch long, the upper lip forming a green hood or beak with red or yellow edges and the tiny lower lip near the center of the tube being green to purple and only about three millimeters long. It can be found on sea bluffs and in dry places in chaparral up to 4,000 feet in most of California. Three subspecies occur, differing in the size and coloring of the flower. Two are found only near the coast, one of them quite rare. The other *(C. affinis* subsp. *affinis)* is more widespread.

Monterey Indian paintbrush

MONTEREY INDIAN PAINTBRUSH ***(Castilleja latifolia)*** is a coastal species with broader bracts and stubbier flowers than the *Castilleja affinis* (see page 151). It is gray green and bristly with wider, more or less fleshy leaves. The bright red to yellow bracts are widely wedge shaped to obovate and usually have three lobes. The calyx is divided about halfway down in the back, but more deeply in the front, and only slightly divided on the sides. The flower is similar to the above species but smaller, with a green upper lip with red edges and a tiny green lower lip. It occurs on coastal sand dunes and in coastal scrub below 500 feet from Monterey County northward. It is not common, but in good years it can be locally abundant.

PURPLE OWL'S-CLOVER ***(Castilleja exserta)*** is in the second group of the *Castilleja* genus and was formerly in the *Orthocarpus* genus. It is a low erect annual with deeply divided leaves and terminal spikes of narrow crimson to purplish or whitish flowers with a lobed white to purple bract below each flower. The tip of the hooded upper lip of the flower curves over, covering the stigma, and it is densely hairy on the back. It

Purple owl's-clover

often occurs in great masses in pastures and on grassy hill-sides and forms patches of great beauty up to about 5,000 feet from Mendocino County to Baja California and inland to the San Joaquin Valley and the deserts. Very similar to it and with a similar range is a plant simply called owl's-clover *(C. densi-flora)*. The tip of the hooded upper lip is erect, not curved over, and usually less hairy on the back.

In the same figwort family is the *Penstemon* genus, commonly called beardtongue because the fifth stamen is sterile and often flattened and bearded like an elongate hairy tongue. A common red species is **SCARLET BUGLER *(Penstemon centranthi-folius)***, a perennial with several waxy stems one to three feet high with thick heart-shaped leaves clasping the stem. The

Scarlet bugler

leaves and flowers are both arranged opposite each other. The tubular scarlet flowers are about an inch long with short erect petal lobes. In this species the sterile stamen is hairless. It is found below 6,000 feet in dry, open or wooded, often disturbed places, generally in chaparral or oak woodland from Lake County southward.

Monkeyflower *(Mimulus)* is another large genus in the figwort family with 63 species in California, all of them native. Many are annual, many are perennial, and the flowers can be yellow, orange, white, pink, red, or purple. **SCARLET MONKEYFLOWER *(Mimulus cardinalis)*** is a freely branched glandular perennial growing up to three feet tall with large, toothed, opposite, light green leaves more or less clasping the stem. The large tubular flowers are bright red to orange with the two-lobed upper lip

Scarlet monkeyflower

arched and the three-lobed lower lip turned downward. It is frequent on stream banks and in wet places up to 8,000 feet in most of California.

A beautiful plant of early spring is **INDIAN WARRIOR *(Pedicularis densiflora)***. It grows from a short woody base up to a height of about two feet bearing a spike of several deep purple red tubular flowers. Each flower is about an inch long and has a hooded upper lip called a galea and a small, inconspicuous lower lip; they are much like the flowers of the *Castilleja* genus described earlier. The large, fernlike leaves are deeply divided into several toothed or lobed segments. It is found below 7,000 feet on dry slopes in chaparral and woodlands throughout

Indian warrior

most of California. A very early bloomer, the flowers may appear as early as January.

California bee plant, or California figwort

CALIFORNIA BEE PLANT, or **CALIFORNIA FIGWORT**, ***(Scrophularia californica)*** is a tall, square-stemmed perennial with opposite, unevenly toothed, oval to triangular leaves. The orange to maroon flowers are small and sometimes easy to miss. About half-an-inch long, they have a short bowl-like tube below, an erect upper lip with two rounded lobes, and three small lower lobes situated along the edge of the bowl. The flower is reported to hold so much nectar in its bowl that bees get drunk from it. It is common in moist places below 5,000 feet throughout California and northward to British Columbia.

WITHERED SNAPDRAGON ***(Antirrhinum multiflorum)*** is in the same genus as the garden snapdragon. It is a stout, widely branched, exceedingly glandular and hairy plant with several rose red to pink snapdragon-like flowers at the ends of the branches. The lower lip of the tubular flower has a white, cream, or tan

Withered snapdragon

withered-looking spot on the lower lip, giving the plant its common name. It is found on dry, rocky slopes, often in disturbed or burned areas, below 6,000 feet from Alameda to San Bernardino Counties and in the central Sierran foothills.

FOXGLOVE *(Digitalis purpurea),* a final reddish member of the figwort family, is a large glandular and densely hairy plant

Foxglove

that can be up six feet tall. The large lanceolate to oval leaves are hairy and have winged stems. The several hairy purple to white flowers are bell shaped, up to two inches long, and usually spotted within. It is a native of Europe but has established itself freely in more or less shaded places below 3,500 feet near the coast from Santa Barbara County to British Columbia. Although the leaves can resemble the herbs comfrey *(Symphytum officinale)* and sage (*Salvia* spp.), the plant is very poisonous to ingest for both humans and livestock.

In the sunflower family (Asteraceae), with its composite heads of several disk flowers, ray flowers, or both, is **VENUS THISTLE *(Cirsium occidentale* var. *venustum)*.** It is a woolly, spiny plant, two to four feet tall, with long, lobed leaves with spiny edges. The flower head is composed of many elongated disk flowers with five long, narrow, petal lobes. It is pollinated by butterflies with their long proboscises that can reach deep into the flower tubes to the nectar at the bottom. The color most often ranges from deep pink to red but can occasionally be white or purple. Below the flower head is a wonderful pattern of spiraled and spine-tipped bracts, or phyllaries, interlaced with a cobwebby white wool. It is found on dry slopes up to 12,000 feet from Mendocino and Butte Counties southward, especially away from the immediate coast. Closely

Venus thistle

Salsify, or oyster plant

related species have more or less wool, vary in stature and color, and have more local distribution. Flowers usually begin in April and continue through May or June.

SALSIFY, or **OYSTER PLANT**, ***(Tragopogon porrifolius)*** is composed of only ray flowers. A single-stemmed plant growing from a very long taproot, it has narrow, elongate light green leaves up to eight inches in length that gradually taper to a long tip and flare out at the base. The flower stem is widened at the top and bears a large round head of many purple ray flowers with toothed, squared-off tips. The phyllaries below the head are long and tapered, generally longer than the ray flowers. In fruit, the head becomes a large dandelion-like puff of feathery hairs with seeds attached below. It occurs in open fields and waste places up to 5,500 feet throughout California. The root is reported to taste like oysters when it is roasted, hence the common name.

BLUISH FLOWERS

Blue to Violet

Blue dicks

One of the more common blue-flowered plants in the lily family (Liliaceae) is **BLUE DICKS** ***(Dichelostemma capitatum),*** with a long leafless stem and two to five narrow leaves about a foot long at the base. The blue, bell-shaped flowers are in dense umbels, that is, the flowers all radiate from one point at the top of the stem. Two purple, papery bracts, or modified leaves, enclose the base of the umbel. The flowers are tubular below and flare out to six lobes above with six stamens, three large with winglike appendages and three smaller with no appendages. This species can sometimes form dense masses of color and occurs in open places in much of California below 8,000 feet. It blooms from March to May.

ITHURIEL'S SPEAR, or **GRASS NUT,** ***(Triteleia laxa)*** is related to blue dicks *(Dichelostemma capitatum),* but the flowers are usually larger, lighter blue, and in more open umbels, each individual flower having a stem from one to four inches long. The six stamens are all equal in length with no appendages, but three are attached lower on the petal lobes and can appear shorter. This species is common in heavy soils below 4,600 feet in the Coast Ranges from Oregon to San Bernardino County and in the Sierran foothills from Tehama to Kern Counties. Flowering is from April to June.

Ithuriel's spear, or grass nut

CAMAS *(Camassia quamash)* is a very attractive plant in the lily family with clusters of beautiful deep blue flowers. It has bulbs up to almost an inch thick, long thin basal leaves up to

Camas

two feet long, and a flowering stalk generally two feet tall but occasionally reaching a height of almost four feet. The flowers are starlike, with six narrow identical sepals and petals up to two inches long and six stamens with anthers ranging from pale yellow to violet. It is found up to 11,000 feet in moist meadows and forests and near streams from Marin and El Dorado Counties northward to Canada. It is especially striking when it forms solid sheets of color in wet meadows, blooming from May to July. The bulbs were an important food source for Native Americans, and battles between tribes were sometimes fought over collecting rights.

The Pacific Coast has several beautiful plants in the iris family (Iridaceae), including those in the iris genus, *Iris*. One of the most common and robust is **DOUGLAS' IRIS *(Iris douglasiana)*,** which forms large clumps of tough fibrous leaves with pinkish or reddish bases and flower stalks generally one to two feet tall. The flowers are the typical form of the well-known garden iris with petals two to three inches long and a narrow one-half to one-inch-long tube below that widens into a long inferior ovary hidden in the large, leaflike bracts below. The flower color varies from pale lavender to blue or deep red purple. It is abundant on grassy slopes and in open places from Santa Barbara County to Oregon.

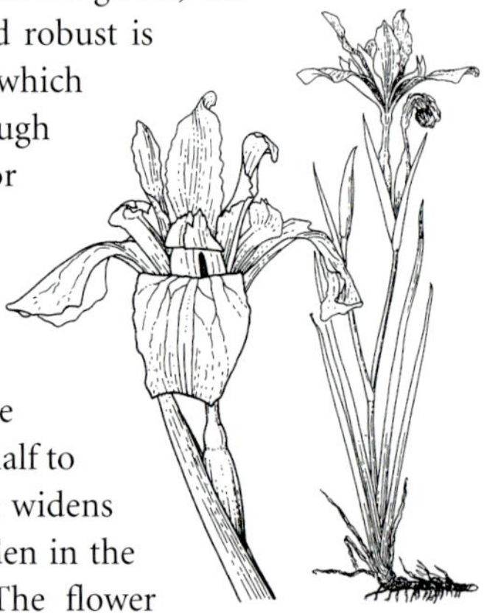

BLUE-EYED-GRASS *(Sisyrinchium bellum)* is not a grass at all but a member of the iris family. The flat, narrow leaves, however, look very grasslike, and the plant can be difficult to distinguish when not in flower or fruit. The blue purple flower, although not the typical iris form, is six parted and tubular

Douglas' iris

Blue-eyed-grass

below and has an inferior ovary, thus identifying it as a member of this family. The center of the flower is a bright shiny yellow, providing a stunning contrast to the brilliant blue and alerting the pollinator to the nectar at the center. It occurs in moist, grassy areas and woodlands below 8,000 feet throughout California.

Yellowtinge larkspur

In the buttercup family (Ranunculaceae) is **YELLOWTINGE LARKSPUR *(Delphinium decorum)*,** in the large larkspur *(Delphinium)* genus, which has almost 30 species in California, all of them native and perennial. A few have white or red flowers, but most are blue flowered, as is this species with two to 20 deep blue purple flowers spread out along the upper stem. Three to 18 inches tall, this plant is generally hairy on the lower part of the stem and has mostly basal leaves divided into three to 15 lobes with somewhat hairy undersides. The flowers have five very showy petal-like sepals, the two upper ones forming half-an-inch long spurs that extend out horizontally behind the flower. In the center are four small hairy petals; the two upper ones are white, and the lower ones are blue. Two subspecies occur, both with their own common names. Coast larkspur *(D. decorum* subsp. *decorum)* is found below 500 feet in grasslands and open chaparral, usually near the coast, from the San Francisco Bay Area northward. It blooms from March to May. Tracy's larkspur *(D. decorum* subsp. *tracyi),* sometimes prostrate and less hairy, occurs above 1,100 feet in meadows in coniferous forests in the Klamath Mountains and northern Coast Ranges and flowers from May to June.

PARRY'S LARKSPUR *(Delphinium parryi)* is a taller plant with flowers ranging from light to dark blue. It can be up to four feet tall with as many as 60 flowers on one plant, although they are generally more spread out than in the above species. The stems and leaves have small, curled hairs, distinguishing it from other hairy species that generally have straighter hairs. It has few, if any, basal leaves, and the stem leaves are divided into five to 27 lobes. It occurs in chaparral and open woodlands below 8,500 feet from western central California southward. Five subspecies occur, differing in the condition of the basal leaves at flowering time, plant and flower sizes,

Parry's larkspur

length of the lower petals, and whether the sepals are spreading or reflexed. Other blue species of larkspur are found in other parts of the state, and several species are known to be poisonous to cattle.

Among the many lupine *(Lupinus)* species in California in the pea family (Fabaceae), one of the more common ones is **ARROYO LUPINE *(Lupinus succulentus)***. Although native, it can be quite weedy and is often found in disturbed places and along highways. It is a fleshy, robust annual about eight to 40 inches tall with a hollow, mostly glabrous, or nonhairy, stem with many long-stemmed leaves divided palmately (like fingers spreading out from the palm of a hand) into seven to nine oblong leaflets. The blue purple flowers are in whorled clusters three to six inches long. The wide upper banner petal has a white patch near the center that turns magenta after the flower has been pollinated, signaling visiting bees that there is no longer any nectar available. The side wing petals on the lower lip are often sparsely hairy, and the keel petal hidden between the wings is slightly to very hairy near the base on both the upper and lower edges. Arroyo lupine is found below

Arroyo lupine

3,000 feet in many types of open habitat, including disturbed and waste places, throughout much of California.

SKY LUPINE *(Lupinus nanus)* is another annual lupine found throughout much of the state. It is at least somewhat hairy and not fleshy, unlike the above species, and the flowers range from sky blue to a deep royal blue. The leaves are smaller,

Sky lupine

more hairy, and on shorter stems than in arroyo lupine *(L. succulentus)*. The flowers are whorled in tighter, more symmetrical clusters, and the individual flowers are smaller with the same white patch on the upper banner petal, but the hidden keel is hairy near the pointed tip, not at the base. It is found below 4,500 feet throughout California and northward to British Columbia in many types of open habitats, often in abundance and forming mass displays of blue that can be seen from quite a distance.

MINIATURE LUPINE *(Lupinus bicolor)* is a third widespread annual lupine. Similar to sky lupine *(L. nanus)*, it is usually a smaller and more delicate plant with narrower leaflets and smaller

Miniature lupine

flowers in shorter, fewer-flowered clusters. A large, robust plant of this species, however, can be very difficult to distinguish from sky lupine. Generally, the stems of the individual flowers are a few millimeters shorter in miniature lupine, and the banner petal is longer than wide, whereas in sky lupine it is as wide as or wider than the length, but these minute differences can be difficult to detect. Both species occur in similar habitats below 5,000 feet throughout the state, but sky lupine often forms dense masses, whereas miniature lupine is usually more scattered and less flamboyant.

A common shrubby lupine is **SILVER LUPINE *(Lupinus albifrons)*,** with densely hairy leaflets covered with long, white, appressed hairs, giving the plant a gray or silvery appearance, especially when seen from a distance. The violet blue to pink flowers are whorled in two- to 12-inch-long clusters. The back of the large upper banner petal is hairy, and the patch near the cen-

Silver lupine

ter is white to yellow, turning purple when the nectar is gone. The boat-shaped keel petal has hairs on the upper edge from near the middle to the tip. Silver lupine grows in sandy or rocky areas up to 6,500 feet throughout California. A very similar shrubby lupine, but limited to coastal areas and having no hairs on the back of the banner petal, is yellow-flowered bush lupine *(L. arboreus),* occurring from central California northward and having either yellow or blue flowers, in spite of its common name.

In the flax family (Linaceae) is a fragile annual, **COMMON FLAX *(Linum usitatissimum)*.** A wispy, smooth-stemmed plant with thin lanceolate leaves, it has five-petaled flowers that are so delicate they can sometimes fall apart at the slightest touch. Each flower has a slender inch-long stem, rounded light blue petals about half-an-inch long, and a pistil with five styles and elongated stigmas. It is found in open, grassy, often disturbed areas below 500 feet throughout California. Also common is small-flowered flax *(L. bienne),* with linear leaves and smaller

Common flax

flowers on shorter stems. It is found in grasslands and woodlands from central California northward to Oregon. Both species have been naturalized here from Europe.

WESTERN MARSH-ROSEMARY, or **WESTERN SEA-LAVENDER,** ***(Limonium californicum),*** in the leadwort family (Plumbaginaceae), is a coastal perennial growing from a woody rhizome (underground stem). It has thick, rough leaves two to six inches long and about two inches wide clustered at the base of the plant. A highly branched flower stem rises above the leaves with many small papery flowers about a quarter-inch long with blue to violet petals and hairy, ribbed calyces (fused sepals) with more or less white lobes. It occurs on beaches, coastal strand, and in salt marshes near the coast throughout California.

BIRD'S EYES ***(Gilia tricolor),*** in the phlox family (Polemoniaceae), is a delicate annual with a spreading branched stem up to almost a foot tall with alternate, finely dissected leaves.

Western marsh-rosemary, or western sea-lavender

Bird's eyes

Blue field gilia

Each flower has a flared yellow tube, five purple spots in the throat at the top of the tube, and five spreading pale to blue violet petal lobes. It occurs on open grassy slopes and plains below 2,000 feet in the Coast Ranges from Humboldt to San Benito Counties, in the Central Valley, and along the base of the Sierra Nevada, especially northward.

Gilia *(Gilia)* is a large genus with many colors. The flowers can be in tight rounded heads like in **BLUE FIELD GILIA *(Gilia capitata)*** or in open clusters like in bird's eyes *(G. tricolor)* described above. Blue field gilia is a variable annual with finely dissected leaves and 50 to 100 small blue five-lobed flowers per head. It occurs in open, usually sandy or rocky areas below 7,000 feet in much of California. It is a spring bloomer and often occurs in large masses. Eight subspecies occur, varying in the degree of hairiness and the size and shape of the individual flowers.

Closely related to gilia *(Gilia)* is **STRAGGLING-GILIA,** or **FALSE-GILIA, *(Allophyllum gilioides),*** a small hairy annual, often quite glandular, with open to dense clusters of two to eight flowers.

Straggling-gilia, or false-gilia

The leaves are deeply divided into very narrow, often linear lobes, and the dark blue purple flowers are less than half-an-inch long. Because of its size, it is easy to overlook, but the lovely color and flower design make it worth finding. It occurs on dry slopes and flats below 6,000 feet, ranging from the foothills of the Sierra Nevada to the mountains of San Diego County.

Another relative of gilia *(Gilia)* is **HOLLY-LEAVED NAVARRETIA *(Navarretia atractyloides)*,** a short glandular-hairy annual with a stout stem and several stiff, deeply lobed, spine-tipped leaves. The midrib of the leaf is about a quarter-inch wide, and the upper lobes are angled forward toward the leaf tip rather than spreading. The leaflike bracts just below the flowers are very broad at the base and have three upward-angled spines at the tip. The blue to purple flowers are about half-an-inch long, the red-veined tube being longer than the tiny petal lobes. It is found in dry, open rocky or sandy places below 1,200 feet in the Coast Ranges from Humboldt County to San Diego and in the Channel Islands. Several other navarretia *(Navarretia)* species are found in the state, many with blue flowers, but a few with white or yellow flowers. Many of them occur in vernal pools, a decreasing habitat in California, but some species are found in drier areas.

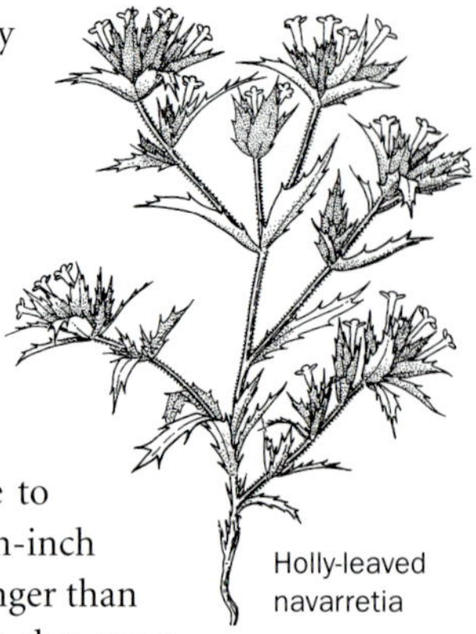

Holly-leaved navarretia

A large and polymorphous genus in the waterleaf family (Hydrophyllaceae) is the phacelia genus *(Phacelia)*, with tight clusters of coiled flower stalks and mostly bell-shaped flowers. **TANSY PHACELIA *(Phacelia tanacetifolia)*** is an attractive, robust, usually branched, glandular annual with stiff rough hairs. The fernlike leaves, up to eight inches long, are divided into several unevenly toothed to lobed leaflets. The dense flower stalks have numerous blue flared flowers with more or less linear, long-haired sepals. The stamens are very long and extend far beyond the flower. The fruit capsule has only one to two seeds, helping to distinguish it from similar-looking species that have four or more seeds per capsule. It is found in open gravel or sandy areas below 7,000 feet from Lake and Butte Counties southward, and westward through the Mojave Desert to Nevada and Arizona.

Tansy phacelia

DOUGLAS' PHACELIA *(Phacelia douglasii)* is another glandular annual, but shorter and with softer hairs. The leaves are mostly basal, long stemmed, and deeply lobed or divided into rounded segments. The light blue to purple flowers are usually larger than those of the tansy phacelia *(P. tanacetifolia)* described earlier. They have shorter sepals and shorter stamens and are arranged much more loosely. Each individual flower has a short stem connecting it to the flower stalk, unlike tansy phacelia, in which the base of each flower is directly attached to the flower

Douglas' phacelia

stalk. The fruit capsule has eight to 20 pitted seeds. Douglas' phacelia occurs in open, mostly sandy areas below 5,500 feet from central California southward and in the western parts of the Mojave Desert. In the central and northern parts of the state, California phacelia *(P. californica)* can look similar, but it is a perennial and has shorter leaves with longer stems, and the fruit has only two seeds.

CALIFORNIA BLUEBELLS *(Phacelia minor)* is a very different-looking phacelia. It is another rough, glandular annual, but the long-stemmed leaves are oval to round and irregularly toothed, not lobed or divided as in the previous species. The striking purple flowers are long tubular rather than flared and up to an inch-and-a-half long with small, more or less spreading petal lobes above and short, narrow, glandular, hairy sepals below. The fruit capsule has 30 to 80 pitted seeds. It is found below 5,000 feet in dry, open areas, often after fires, and on disturbed sites in much of southern California. It flowers from March to June.

California bluebells

BABY BLUE-EYES ***(Nemophila menziesii)*** is also in the waterleaf family, but instead of having its flowers on coiled stems, it is recognized as a member of this family by the small, reflexed, sepal-like appendages occurring between the sepals. It is a delicate, sprawling spring bloomer with opposite leaves divided into toothed leaflets. The flower is usually a lovely shade of baby blue, often with black dots and lines decorating the

Baby blue-eyes

petals. It ranges in size from one-quarter to one inch long and one-quarter to one-and-a-half inches wide. It is found in grassy and brushy places up to about 6,000 feet in much of California and blooms from February to June. A white-flowered variety known commonly as white baby blue-eyes *(N. menziesii* var. *atomaria)* grows near the coast from central California northward to Oregon.

Fiesta flower

Related to baby blue-eyes *(Nemophila menziesii)* and having the same characteristic reflexed appendages between the sepals is **FIESTA FLOWER *(Pholistoma auritum)***. A sprawling annual with recurved prickles on the stem, it clings to nearby vegetation for its support, as well as to the clothing of unsuspecting hikers. It has an interesting deeply lobed leaf with a widely winged stem that expands at the base to reach around and clasp the central stem. The flowers are about half-an-inch long and up to an inch wide or more. It grows in deep canyons and on shaded slopes below 6,000 feet from Lake and Calaveras Counties southward, blooming from March to May.

Pholisma

Along the sandy beaches of California is a small fleshy herb, **PHOLISMA** ***(Pholisma arenarium),*** of the lennoa family (Lennoaceae), a small family of parasitic plants depending on the roots of other plants for their nourishment because they lack chlorophyll of their own. It is about four to eight inches tall, and the white stem, turning brown in age, has small, glandular, scalelike leaves. The several purple flowers with white borders are each about an eighth-of-an-inch wide. It is found in sandy soil on coastal dunes from San Luis Obispo County southward and in the deserts of southeastern California, Arizona, and northwestern Mexico.

PURPLE NIGHTSHADE ***(Solanum xanti)*** is a close relative of potato, tomato, eggplant, and green pepper, all in the nightshade family (Solanaceae), and all having stamens with anthers longer than the filaments. Purple nightshade is a branched, more or less hairy shrub growing up to three feet tall. The more or less flat open flower, half-an-inch to an inch in diameter, is dark blue to lavender with bright yellow anthers in the center. The

Purple nightshade

berrylike fruit resembles a small green tomato. It is found in shrubby and wooded areas below 9,000 feet throughout California. Also common in most of the state, except the northwest, is blue witch *(S. umbelliferum)*, a more densely hairy plant with at least some forked hairs. The flowers range from lavender to deep purple with green and white markings at the base and the same large bright yellow anthers in the center. The round fruit turns purple when ripe. A rare species of nightshade, Wallace's nightshade *(S. wallacei)* with larger flowers and yellowish fruits, occurs in the Channel Islands.

The mint family (Lamiaceae), with its aromatic qualities, paired leaves, and two-lipped flowers, has several blue-

Dannie's skullcap

flowered plants such as **DANNIE'S SKULLCAP** ***(Scutellaria tuberosa),*** a perennial with thin underground rhizomes producing small tubers. Less than 10 inches tall, it has a hairy stem and oval leaves with more or less rounded teeth. The sepals are fused into a tube, or calyx, with a crestlike projection above, giving the plant its common name. The tubular two-lipped flowers, about half-an-inch long, are deep violet blue with white markings on the lower lip. It is common between 500 and 3,500 feet along the edges of chaparral and oak woodlands, especially after fires, throughout most of California.

WOOLLY BLUECURLS ***(Trichostema lanatum)*** is a low, rounded shrub, generally between two and four feet tall, with crowded, aromatic, lanceolate leaves, green above but gray hairy below. The plant is covered with long, densely hairy spikes of many erect inch-long blue flowers with very long arching stamens that can be almost two inches long and extend far beyond the flower. The sepals are fused into a tube, or calyx, and densely covered with very woolly pink, blue, purple, or white hairs,

Woolly bluecurls

which provide much of the plant's color. It can be found below 3,500 feet in chaparral and coastal scrub from Monterey County to Baja California. It has been used as an astringent and also for sores and ulcers.

Sage *(Salvia)* is a wonderfully fragrant genus in the mint family, and several species are commonly used in cooking. Most sages are perennial, but **CHIA *(Salvia columbariae)*** is one of only two annual species in California. About four to 20 inches tall, it has rough leaves deeply lobed into toothed or lobed segments. The upper part of the

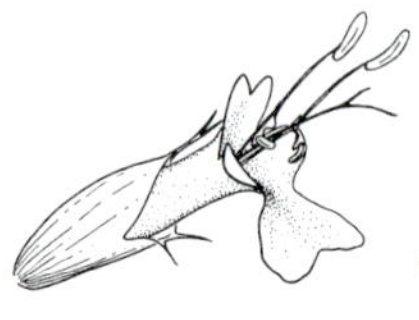

Chia

stem has several tiers of whorled, deep blue, two-lipped flowers with purple to deep rose calyces, forming an exquisite color combination. It is common on dry sites in chaparral and coastal-sage scrub below 4,000 feet from inner Mendocino County southward to Mexico. The seed is very nutritious and was used extensively by the Native Americans and early pioneers to make a refreshing drink. It was also roasted and used in many foods.

THISTLE SAGE *(Salvia carduacea)* is California's other annual species of salvia. It is a very spiny, white-woolly plant with mostly basal, thistlelike, lobed leaves with wavy, spiny edges. The whorls of lavender flowers have spiny bracts below. The sepals are also woolly and spine tipped, and the petal lobes of the inch-long flowers are fringed on the edges. The long stamens extend far beyond the flower tube and have bright red anthers. Thistle sage occurs below 4,500 feet in sandy and gravelly places of the interior from the eastern San Francisco Bay Area southward to Baja California.

Thistle sage

One of the many perennial sages in California is **BLUE SAGE** ***(Salvia clevelandii)***, a fragrant, hairy shrub growing up to about three feet tall with puckered ashy gray leaves. The flowers are in one to three whorls on the stem with several firm dark bracts below each whorl. The two-lipped flowers are dark blue violet and almost an inch long, with four long stamens and a pistil extending far beyond the flower tube. It is found below 3,500 feet on dry slopes in chaparral and coastal scrub from San Diego County southward to northern Baja California. The dried leaves, if used sparingly, can be used in cooking.

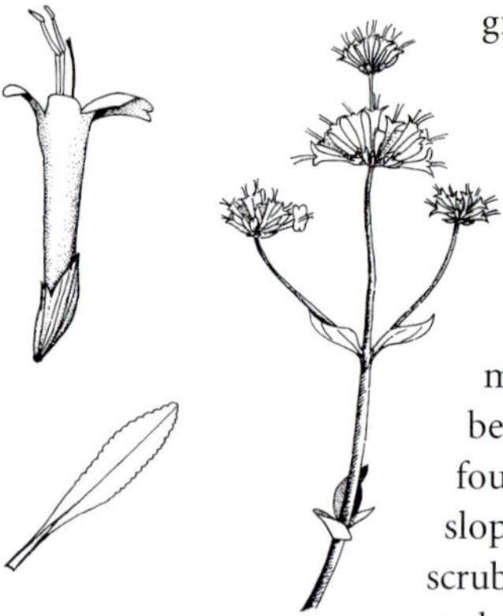

Foothill penstemon

FOOTHILL PENSTEMON ***(Penstemon heterophyllus),*** in the figwort family (Scrophulariaceae), is one of the most widespread members of the beardtongue genus, *Penstemon,* in California, which is characterized by having four pollen-producing stamens and a fifth sterile stamen. Somewhat woody at the base and branched, this species is generally about one to one-and-a-half feet tall but can occasionally grow to almost five feet in height. The narrow leaves are one to three inches long with a tapered base, and the flowers are about an inch long with a long, thick tube opening up into smaller petal lobes above. While still in bud the flowers appear to be yellow, but upon opening they range in color from magenta to blue. Foothill penstemon grows on dry hillsides below 5,000 feet in most of the state and begins to flower in April. A closely related species found in similar habitats but limited mostly to northern California and the Sierra Nevada is azure penstemon *(P. azureus)*, with wider leaves and flowers that appear blue in bud as well as in flower. Both species have four stamens with hairy anthers and a glabrous fifth sterile stamen, unlike many species in this genus in which the sterile stamen is quite hairy.

NOTABLE PENSTEMON *(Penstemon spectabilis)* is one of the most conspicuous species of *Penstemon* in California. It is a large plant generally three to four feet high, woody below and

Notable penstemon

branched from the base. The lower leaves are long and thin, but the upper are much larger and wider. Strongly toothed and opposite each other, they are more or less triangular with their bases fused together and surrounding the stem. The blue to blue purple tubular flowers are about an inch long and whitish inside. The fifth sterile stamen is glabrous. This species is found below 8,000 feet on gravelly and sandy slopes and along washes in chaparral, coastal-sage scrub, and oak woodland from Los Angeles County southward to Mexico.

Another plant in the figwort family is **CHINESE HOUSES** ***(Collinsia heterophylla)***, an annual with opposite lanceolate to triangular leaves, each pair situated at a 90-degree angle to the pairs immediately above and below. The flowers, white to pale blue on the upper lip and darker blue to purple on the lower lip, are whorled around the stem in progressively smaller lay-

Chinese houses

ers or tiers, like a Chinese pagoda. The individual flowers resemble those of the pea family, but the two side wing petals of the lower lip are connected to the keel petal below, and if one gently pulls sideways on the wings, the keel opens and reveals the stamens and pistil. In the pea family, the wing and keel petals are separate. Chinese houses can be found in several different kinds of shady habitats below 3,000 feet throughout California, often in large colorful masses.

FEW-FLOWERED BLUE-EYED MARY ***(Collinsia sparsiflora)*** is a shorter, more open and delicate annual in the same genus with small, individual, long-stemmed flowers spaced out along the stem. The leaves are narrow and opposite one

Few-flowered blue-eyed Mary

another. A long, thin flower stem arises from the leaf pair and bears a single lavender, blue, or purple flower the same shape as those of the above species but much smaller and often hairy on the lower lip. Occurring up to 4,000 feet in rocky and stony places in grasslands, woodlands, and chaparral, it can be found from central California northward.

California vervain, or western verbena

CALIFORNIA VERVAIN, or **WESTERN VERBENA,** ***(Verbena lasiostachys)*** is a coarse perennial, one to three feet tall, in the vervain family (Verbenaceae). The square stem is covered with short, spreading, generally rough hairs. The toothed oval leaves are opposite each other on the stem and can be soft to very rough hairy. The numerous small blue flowers are on spikes three to 10 inches long at the top of the plant. Each of the tubular blue flowers is only about a quarter-inch long, with soft or rough hairy sepals. It is found in open, often wet areas up to about 8,000 feet throughout the state. In southern California a similar species is mint-leaved verbena *(V. menthifolia),* with very stiff, appressed hairs on the stems, leaves, and sepals.

Jim brush

JIM BRUSH *(Ceanothus oliganthus* var. *sorediatus)* is a very attractive shrub in the buckthorn family (Rhamnaceae). It can be up to about 10 feet tall and is sometimes treelike and very branched with alternate evergreen leaves with three central veins. The many small light to deep blue flowers are in large clusters about an inch-and-a-half long. It occurs on dry, shrubby slopes below 4,000 feet in the Coast Ranges and in the Transverse and Peninsular Ranges of southern California. Also found in southern California is hairy ceanothus *(C. oliganthus* var. *oliganthus),* with hairy twigs and leaves and darker blue or purple flowers. Many species of ceanothus *(Ceanothus)* occur in California, and several are used in cultivation, often found planted in parks. The genus is often called California-lilac, and the flowers of the various species range from white to a deep blue purple.

In the bellflower family (Campanulaceae) the sepals and petals are fused into tubes that spread out into four or five lobes above, and the ovaries are inferior, that is, they occur below the sepals and petals instead of above. **COMMON BLUE-CUP *(Githopsis specularioides)*** is a small branched annual generally only a few inches tall with delicate stems and few small, toothed leaves. The inconspicuous dark blue flowers have bell-shaped inferior ovaries and one to several toothed bracts situated below the flowers. Common bluecup is found in chaparral and oak woodlands below 4,500 feet in many parts of California.

Common bluecup

Cuspidate downingia

CUSPIDATE DOWNINGIA *(Downingia cuspidata)* is another, although different-looking, member of the bellflower family, very similar to the popular garden plant, lobelia (*Lobelia* spp.). A small annual, it has narrow leaves that are often deciduous before flowering. The very short flower tube spreads out into a small upper lip deeply divided into two narrow spreading blue lobes and a larger, shallowly three-lobed lower lip, pale to bright blue or lavender with a white center and two yellow spots near the throat. The inferior ovary is long and stemlike. It is found below 1,500 feet in wet places such as vernal pools, meadows, and along the edges of lakes in many parts of California. Several other species of *Downingia* occur in the state, distinguished from one another by the shapes of the upper petal lobes, the color and markings on the lower lip, and small details in the anthers.

SEASIDE DAISY *(Erigeron glaucus)* is a strictly coastal plant in the sunflower family (Asteraceae) with its heads of petal-like ray flowers and small tubular disk flowers. A low perennial with

Seaside daisy

fleshy, more or less spoon-shaped leaves about three to six inches long, seaside daisy can be glabrous, glandular, or densely hairy. The flower heads are made up of numerous yellow disk flowers surrounded by as many as 165 ray flowers about half-an-inch long and ranging in color from blue to lavender to white. Each plant can have from one to 15 flower heads, and each head is about one to two inches wide. Seaside daisy occurs on coastal beaches, dunes, and bluffs the length of the state, as well as in the northern Channel Islands, and flowers from April to August.

YELLOWISH FLOWERS

Greenish Yellow or Yellow to Orange

Leopard lily

LEOPARD LILY *(Lilium pardalinum)* is a tall, elegant member of the lily family (Liliaceae) and one of the most widespread members of the lily genus, *Lilium*, in California. It is a stout plant with several whorls of long narrow leaves along a stem that can be up to nine feet tall. The inverted hanging flowers, in clusters of one to 35, range from pale orange to red with maroon spots outlined in yellow or orange. Each flower can be more than four inches long, and they are often paler on the inside, becoming darker toward the tips of the perianth segments (identical petals and sepals). The large anthers are magenta, orange, or yellow with reddish brown to yellow pollen. Leopard lily occurs in various moist habitats and along stream banks below 7,000 feet. There are five subspecies, some of them quite rare and very geographically limited, but *L. p.* subsp. *pardalinum* occurs throughout most of the state. A similar species but of drier habitats and found only in southern California is ocellated lily *(L. humboldtii* subsp. *ocellatum)*. At one time it was a very common plant in yellow pine forests, but it has declined sharply over the years and is now rarely encountered.

YELLOW MARIPOSA *(Calochortus luteus)* is a bowl-shaped member of the beautiful *Calochortus* genus in the lily family. The slender stem is from eight to 20 inches tall with mostly

Yellow mariposa

basal leaves and one to seven flowers per plant. The three sepals are long and narrow, and the three bright yellow petals are wider, being about one to two inches wide and long. Each petal has red markings below and an oblong to crescent-shaped nectary covered with short hairs. A red brown blotch often occurs near the center of each petal. Yellow mariposa occurs in heavy soils in grassland and wooded areas below 1,100 feet from central California northward and in the Channel Islands.

COMMON GOLDENSTAR ***(Bloomeria crocea)*** has a simple stem with a single long, linear basal leaf that is usually withered by the time the flowers bloom. The top of the stem holds a loose umbel of 10 to 35 or more golden yellow flowers, each consisting of six spreading identical sepals and petals about half-

Common goldenstar

an-inch long with light brown stripes, and six stamens with cuplike appendages at their bases. It can be found below 6,000 feet in grasslands, open woodlands, and along the edges of chaparral from Monterey County to Baja California, flowering from April to June.

GOLDEN BRODIAEA, or **PRETTY FACE**, ***(Triteleia ixioides)*** may be confused with common goldenstar *(Bloomeria crocea)*, described above, as it has similar basal linear leaves and a simple stem with a loose cluster of many flowers at the summit. However, there is usually more than one leaf, the flowers are mostly a paler, less golden, yellow, and the sepals and petals are more oval and form a short tube below. The six stamens are much different, being unequal in length with wide filaments (stamen stems) and two pointed forked appendages behind the anthers. It occurs mostly in sandy soils in coniferous and mixed forests below 10,000 feet in the Sierra Nevada,

Golden brodiaea, or pretty face

the Coast Ranges from San Luis Obispo north, Klamath Mountains, and Cascade Range, and into Oregon. Flowering begins in March and continues into summer.

Golden-eyed-grass

In the iris family (Iridaceae) is **GOLDEN-EYED-GRASS *(Sisyrinchium californicum)***, a plant that can easily be mistaken for a grass when not in flower, with its narrow, flat, very grasslike leaves. When in bloom, however, the bright yellow flowers clearly indicate that it is not a grass, although it does not have the typical flower of the garden iris, either. Rather, it can be identified as a member of the iris family by the six perianth parts, the tubular shape below, and the inferior ovary enclosed in a pair of leaflike bracts below the flower. The open, wheel-like flowers are about an inch in diameter and have six spreading brown-veined yellow perianth segments and three stamens. It is found in moist places below 2,000 feet near the coast from central California to British Columbia.

YELLOW SKUNK-CABBAGE *(Lysichiton americanum),* related to the cultivated calla-lily *(Zantedeschia aethiopica),* is in the large, mostly tropical arum family (Araceae). It has a large, single yellow petal-like bract, or spathe, up to eight inches long cupped or wrapped around a fleshy central stalk known as a spadix that bears numerous small greenish yellow, four-parted flowers with a very unpleasant odor that attracts insects for pollination and gives the plant its common name.

Yellow skunk-cabbage

Later in spring, large thick leaves from one to five feet long appear on short broad stems, and the flowers produce greenish white, berrylike fruits. It is found in marshy areas and along streams below 4,000 feet, generally in coniferous forests from the Santa Cruz Mountains to Del Norte County and on to Alaska and Montana.

Cow-lily, or yellow pond-lily

The southern parts of California have so few natural bodies of water that native water-lilies would seem very unlikely, but **COW-LILY**, or **YELLOW POND-LILY**, ***(Nuphar luteum*** **subsp.** ***polysepalum)*** can be found from San Luis Obispo and Mariposa Counties northward. In the water-lily family (Nymphaeaceae), it has oblong to oval floating leaves four to 16 inches wide with deeply indented bases. The yellow to red-tinged flowers are about two to three inches wide with two layers of large, petal-like sepals, the outer generally green, the inner yellow, and 10 to 20 narrow stamenlike yellow petals. The numerous stamens are also yellow, and the pistil has a broad, disklike stigma. It is found in ponds and sluggish streams below 8,000 feet and begins to flower in April.

California buttercup

In the buttercup, or crowsfoot, family (Ranunculaceae), the most common and widespread buttercup in the state is **CALIFORNIA BUTTERCUP *(Ranunculus californicus)***. A perennial herb growing from five inches to more than two feet tall, it has bright green, sometimes hairy leaves deeply lobed into three to five segments. The flowers have seven or more shiny yellow petals, many stamens, and several small pistils, each of which matures into a round, flat, one-seeded dry fruit with a small hook at the tip. It can be found on moist grassy and wooded slopes below 8,000 feet throughout the state and often begins to bloom as early as February. A few very similar-looking species of buttercup have less than seven petals or slightly larger fruits. The genus name comes from the Latin for "little frog," referring to the fact that this flower occurs in the same moist habitats where frogs abound.

BUSH POPPY *(Dendromecon rigida)* is a yellow-flowered shrub in the poppy family (Papaveraceae). Growing from three to almost 10 feet tall, it has grayish, narrow willowlike leaves slightly toothed on the edges. The silky, bright yellow flowers

Bush poppy

have four petals about an inch long, several stamens, and a more or less flat-topped pistil. It is found on dry slopes and in washes below 6,000 feet, especially after fires, in much of the state. A similar species, Channel Island tree poppy *(D. harfordii)*, is quite rare, found only in the Channel Islands, and has shorter leaves with rounded tips and no teeth on the edges.

CALIFORNIA POPPY *(Eschscholzia californica)* is probably the best known and most widespread wildflower in California, as well as being the state flower. It is highly variable and can be an annual or a perennial, erect or spreading, and short or tall, ranging from only five inches to almost two feet in height. The stems and multidissected leaves often have a waxy film, giving the plant a gray green appearance. The two protective sepals are fused into an erect, pointed tube that splits and pops off when the flower opens. The four petals range in color from orange (inland form) to yellow (coastal form) and can each

California poppy

be more than two inches long. A flat pink disk, or torus, sits below the flower, which is an identifying characteristic of this species. It occurs in grassy areas throughout the state up to 7,000 feet. It begins to bloom as early as February and can often still be found blooming in October. The genus is named after a nineteenth century Russian naturalist.

WIND POPPY *(Stylomecon heterophylla)* is an erect, usually several-stemmed annual one to two feet tall with lobed leaves and terminal four-petaled flowers. The orange red petals become purplish near the base, contrasting with the numerous yellow stamens and green pistil. The petals are so delicate that even the slightest breeze or touch can sometimes cause them to fall off. It is found on grassy and brushy slopes and in chaparral openings below 4,000 feet from Lake County through the Coast Ranges to northern Baja California, as well as in the San Joaquin Valley and the southern Sierra Nevada. It can be distinguished from the somewhat similar-looking fire poppy *(Papaver californicum)*, which often grows with it, by its oranger petals, a clear rather than milky

Wind poppy

sap, and a more spherical and elevated stigma above the seedpod, whereas fire poppy has a flatter stigma that sits directly on the top of the seedpod.

GOLDEN EAR-DROPS *(Dicentra chrysantha)* is a yellow-flowered relative of the pink bleeding-heart *(Dicentra spectabilis)* familiar to gardeners and people from the East Coast. Formerly in the fumitory family, but now a member of the poppy family, it is a tall, coarse perennial up to six feet high with large, very

Golden
ear-drops

dissected leaves. The unusually shaped flowers are about half-an-inch long with four yellow petals, the outer two pouched below and spreading above, and the inner two erect and connected at the top. It usually appears in profusion after a fire on dry slopes and in other disturbed areas below 6,000 feet in most of the state and into Baja California.

A very handsome shrub with large, soft, hairy yellow flowers is **FREMONTIA**, or **FLANNELBUSH**, ***(Fremontodendron californicum)***, in the cacao family (Sterculiaceae), a mostly tropical family with only three native species in California. Frequently used in landscaping and along highways, flannelbush is generally a tall shrub branched from the base. The en-

Fremontia, or flannelbush

tire plant, including stems, leaves, and flowers, is sparsely to densely covered with small, branched, starlike hairs. The wide leaves are generally shallowly lobed and usually more densely hairy on the undersides. The open flat flowers, one to two or more inches in diameter, have no petals, but the five large silky sepals are very petal-like, being bright yellow and often edged in red or orange. It is found in various forms between 1,000 and 6,500 feet in chaparral, oak, and pine woodlands, and rocky ridges in the Sierran foothills and inner Coast Ranges from Lake and Napa Counties southward, as well as in scattered stations in southern California and Arizona. Flowering is in April and May.

Very early in the year, **FIELD MUSTARD**, or **TURNIP**, ***(Brassica rapa)*** starts to blanket vast areas of the state with its bright yellow flowers. Introduced from Europe, it is a widespread, very invasive weed of the mustard family (Brassicaceae) with large, lobed leaves and four-petaled yellow flowers that produce long, narrow seedpods. The lower leaves are typical mustard leaves with two to four side lobes and a large rounded terminal lobe, but the upper leaves are smaller, usually triangular and unlobed, and their bases clasp around the central stem. The flowers have four bright yellow petals, each about half-an-inch long or less. The seedpods are one to three inches long with a long beak ranging from one-quarter to one inch long. It occurs in fields, orchards, and other disturbed areas below 5,000 feet throughout California and much of the rest of the United States. Several other mustards are also widespread and invasive, but most of them have smaller, paler yellow flowers and lack the longer fruit beak and the upper clasping leaves.

Field mustard, or turnip

WESTERN WALLFLOWER *(Erysimum capitatum)* is a native member of the mustard family and less widespread than field mustard *(Brassica rapa),* described above, although still fairly common. It is at least somewhat hairy with multibranched hairs. The long narrow leaves, very different from those of field mustard, occur in a basal rosette as well as along the stem. The four-petaled flowers range

Western wallflower

from yellow to orange and are often densely clustered at the top of the stem, producing very long, thin fruits one to six inches long but only a few millimeters wide. It is found up to 13,000 feet in many different habitats, usually inland, throughout the state. There are four subspecies, two of them rare and geographically limited, and one occurring only at high altitudes in the mountains. The fourth *(E. capitatum* subsp. *capitatum),* the one illustrated, is widespread.

Bladderpod

Another family with four-petaled flowers is the caper family (Capparaceae), known mostly for the caper used in cooking. One of the few California natives in this family is **BLADDERPOD** ***(Isomeris arborea),*** a low, very branched shrub with unpleasant-smelling leaves divided into three leaflets. The terminal flower clusters bear several yellow flowers with long stamens, and the fruits are highly inflated round pods with long stems. It is found in saline, sandy places such as coastal bluffs and desert washes below 4,000 feet from central California southward to Baja California, as well as in the deserts and the Channel Islands.

In the cactus family (Cactaceae) is **STRAWBERRY CACTUS** ***(Mammillaria dioica),*** a short, squat plant ranging from only a few inches to a foot high. It is covered with bumpy tubercles tipped with hairy clusters of spreading white spines. In the

Strawberry cactus

center of each cluster are one to four erect broader-based spines with hooked tips. The yellow to white flowers are about an inch long and one to two inches wide, and the club-shaped fruit is scarlet. It occurs in sandy soils below 5,000 feet in southern California and in parts of the Colorado Desert, blooming from February to April.

MESA TUNA, or **VASEY'S PRICKLY-PEAR**, ***(Opuntia × vaseyi)*** is one of many hybrid (indicated by the × in the name) cacti in the *Opuntia* genus. The stem is made up of flattened joints with clusters of long spines and stiff prickly hairs called glochids.

The flowers are yellow, orange, or red, and they have orange yellow stamens and a pink pistil with a green stigma. The fleshy red purple fruit is juicy and quite tasty, but you must be certain to remove all the glochids from the outer surface or your tongue will suffer. It is found in chaparral and disturbed areas below 1,000 feet in southern California and flowers in May and June.

The stonecrop family (Crassulaceae) also has fleshy succulent plants but without spines. In the liveforever genus, *Dudleya,* is **SEA-LETTUCE *(Dudleya caespitosa)***. It is branched from the base and has several basal rosettes of succulent oblong to lanceolate

Sea-lettuce

leaves about a quarter-inch thick. The flower stems curve up from the outer edges of the rosette and have terminal clusters of small yellow flowers with short triangular sepals and five longer, lanceolate, overlapping yellow petals. It can be found on sea bluffs from central California southward and also in the northern Channel Islands. A plant often associated with it is powdery dudleya *(D. farinosa)*, with grayer, broader leaves and paler yellow flowers. Both species flower from April to July.

YELLOW SAND-VERBENA ***(Abronia latifolia),*** in the four o'clock family (Nyctaginaceae), is a large perennial plant forming mats several feet across. The leaves are round, fleshy, and often glandular, and the yellow flowers are in tight clusters with 17 to 34 flowers per cluster. Each flower has a narrow, half-an-inch long tube below, five spreading petal lobes above, four or five stamens within, and a pistil with a linear stigma at the top. This species ranges from Santa Barbara County to British Columbia and blooms from May to October. It can hybridize

Yellow sand-verbena

Golden currant

with the two pink coastal species in the genus, beach sand-verbena *(A. umbellata)* and red sand-verbena *(A. maritima)*, and a close search of an area usually turns up some plants intermediate in their characteristics.

The gooseberry family (Grossulariaceae) has only one genus, *Ribes*, and its members are all shrubs. Most have pink or red flowers, but **GOLDEN CURRANT *(Ribes aureum)*** is one of the few with yellow flowers. It is a spineless shrub, generally less than 10 feet tall, with light green lobed to toothed leaves with wedged to heart-shaped bases. The inverted, hanging flowers are in clusters of five to 15, and each flower has five spreading yellow sepals and five smaller erect petals. The smooth fruit produced by the inferior ovary is red, orange, or black, and about a quarter-inch long. It can be found in many different habitats in most parts of California below 10,000 feet.

Pacific silverweed

PACIFIC SILVERWEED ***(Potentilla anserina*** **subsp.** ***pacifica),*** in the rose family (Rosaceae), is another coastal plant. The leaves are pinnately compound, that is the leaflets are borne to the two sides of a central axis, and the 10 to 20 main leaflets are bright green above and densely gray hairy underneath. Very small reduced leaflets are usually present between the main leaflets. The solitary five-petaled bright yellow flowers are about an inch wide. It can be found along coastal strand and in salt marshes from Los Angeles County to Alaska and blooms from April to August.

Also in the potentilla *(Potentilla)* genus, but usually found more inland, is **STICKY POTENTILLA** ***(Potentilla glandulosa),*** a loosely branched perennial, also with pinnately compound leaves as in the previous species, but usually more rough haired and a duller green with viscid, or sticky, glands on the undersides. The flowers are pale yellow to cream with five petals about half-an-inch long or less, five sepals, and five sepal-like bracts. It is common in brushy places and canyons up to 12,500 feet throughout California, flowering from May to July. There are eight subspecies varying in hairiness, glandularity, style length and thickness, and petal and leaflet details.

Sticky potentilla

In the pea family (Fabaceae), the lotus genus *(Lotus)* has many yellow-flowered plants such as **LARGE-FLOWERED LOTUS *(Lotus grandiflorus)***. Growing one to two feet high, it has leaves made up of seven to nine leaflets arranged pinnately (leaflets spreading to either side of the midrib). The yellow flowers, aging red, are the typical pea family form with a wide upper banner petal, two side wing petals, and a boatlike keel petal hidden between the wings. Almost an inch long, they are arranged in tight clusters of three to nine flowers at the top of a flower stalk that can be more than two

Large-flowered lotus

inches long. This species grows on dry slopes below 6,000 feet, mostly in brushy places in much of California.

BIRDFOOT TREFOIL ***(Lotus corniculatus)*** is a very common bright yellow nonnative weed found in moist places through much of the United States. The several stems are generally decumbent (the lower parts prostrate, the upper erect), and they

Birdfoot trefoil

spread out from a central point. The compound leaves are divided into five small leaflets, two near the base of the midrib and three further up. The three to eight yellow pea-shaped flowers, about half-an-inch long each, are in a cluster at the top of a long flower stem that arises from the leaf axil. It can be found in moist disturbed places almost anywhere in California below 3,500 feet.

Woolly trefoil

WOOLLY TREFOIL ***(Lotus humistratus)*** is a common native lotus that, in spite of its name, is sometimes only slightly hairy. More often, however, it is densely hairy with a grayish appearance. Usually only a few inches tall, it can be prostrate, decumbent, or erect with leaves divided into four small leaflets. A single stemless yellow pea-shaped flower, about half-an-inch long or less, sits in the leaf axil. The sepals are fused into a calyx consisting of a short tube and five spreading pointed lobes that are usually longer than the tube. Very similar and sometimes almost impossible to distinguish is common trefoil *(L. wrangelianus)*, usually not hairy or only slightly so, with calyx lobes equal to or shorter than the tube. Both species can be found in many habitats below 5,000 feet throughout the state and often grow together, but woolly trefoil usually favors drier habitats.

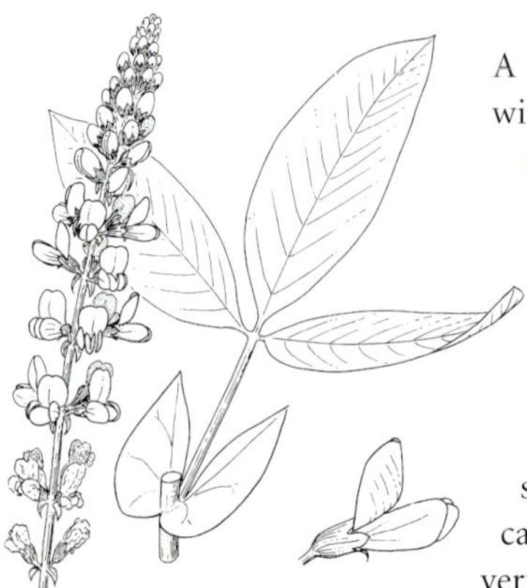

A member of the pea family with larger yellow flowers is **FALSE-LUPINE**, or **GOLDEN-PEA**, ***(Thermopsis macrophylla),*** a stout perennial herb with compound leaves divided into three rounded leaflets from one to more than two inches long. Ranging from slightly to very hairy, the plant can appear green, gray, or silvery. The bright yellow flowers, up to almost an inch long, differ from plants in the true lupine genus in having 10 free stamens, whereas in the lupine genus

False-lupine, or golden-pea

Bermuda-buttercup

the filaments of the stamens are fused. There are four subspecies, two of them rare, varying in the degree and type of hairiness and the density of the flower clusters. The species grows below 7,000 feet through much of California and into Oregon.

BERMUDA-BUTTERCUP *(Oxalis pes-caprae)* is a ubiquitous and pesky weed from Africa. In the oxalis family (Oxalidaceae), it grows from bulbs and has several cloverlike leaves on long stems growing from near the base of the plant. The bright yellow inch-wide flowers are on stems up to a foot long, and the five overlapping petals form a bowl below and spread out

above. There are generally 10 stamens, five short and five long. The plant spreads rapidly by underground stems and bulblets and often becomes a rampant pest almost impossible to get rid of in lawns, parks, and other disturbed areas below 2,000 feet. It starts to grow as early as November and can sometimes still be found blooming into the summer months. Children know this plant as sour-grass and love to suck on the stem with its tart, sweet-sour acidic sap. Too much, however, can cause a stomach ache, and very large amounts can be toxic.

Common meadowfoam

One of the most distinctively Californian wildflowers is **COMMON MEADOWFOAM** ***(Limnanthes douglasii),*** a low-growing branched annual in the meadowfoam family (Limnanthaceae) with leaves dissected into five to 13 leaflets. The delicate, more or less cup-shaped flowers can be yellow, white, or yellow with white tips, with fine purple, pink, or cream veins. The petals are half-an-inch to almost an inch long and become reflexed as the fruit matures, whereas in a few similar species the drying petals curve in toward the center of the flower during fruiting. It often grows in great masses in moist places below 3,500 feet, particularly in the inner Coast Ranges from San Luis Obispo County to southern Oregon. There are four subspecies varying in the coloring of the petals and veins. Yellowish forms appear nearer the coast and pinkish ones in the Sierran foothills. Related species are found as far south as the Laguna Mountains of San Diego County.

The violet family (Violaceae) includes the popular garden pansy *(Viola wittrockiana)*, as well as 21 native California species varying widely in color and leaf characters. One of the yellow-flowered violets is **SHELTON'S VIOLET *(Viola sheltonii)*,** with large leaves up to almost three inches long divided into many linear lobes. The deep lemon yellow flowers consist of two erect upper petals with brownish purple backs, two spreading side petals with club-shaped hairs, and a wider lower petal with brown purple veins. It grows in brushy or more or less wooded places in rich or gravelly soil from 2,500

Shelton's violet

Johnny-jump-up

to 8,000 feet, from Orange County northward to Washington. It begins flowering in April.

A large-flowered yellow violet is **JOHNNY-JUMP-UP** ***(Viola pedunculata)***. It grows two to 15 inches high from a spongy rhizome (underground stem) with many fleshy roots. The stem is often branched and somewhat hairy with several long-stemmed, more or less roundly triangular, toothed leaves. The flowers are orange yellow with two upper petals with red brown backs, two spreading hairy side petals, and a wider lower petal with dark brown veins. It occurs on grassy slopes below 2,500 feet, from Sonoma and Colusa Counties to Baja California.

The elegant yellow flowers of **SAN JOAQUIN BLAZING STAR** ***(Mentzelia pectinata)***, in the loasa family *(Loasaceae)*, belie a very rough plant covered with sharp barbed hairs. A branched annual, it grows from three to 21 inches tall with unevenly lobed leaves up to five inches long. The bright yellow flowers

have five petals, each up to an inch long and often having a red or orange spot at the base. A narrow, generally toothed bract sits below each flower. The multiple stamens are usually of uneven lengths, and the inferior ovary situated below the sepals and petals is long and more or less cylindrical. This species is usually found on steep slopes in grasslands and oak woodlands up to 6,000 feet in the southern Sierra Nevada, Tehachapi Mountains, and inner South Coast Ranges.

A larger-flowered and even more striking member of the blazing star genus, *Mentzelia,* is **LINDLEY'S BLAZING STAR *(Mentzelia lindleyi),*** with petals from one to two inches long

Lindley's blazing star

and about an inch wide. Similar to San Joaquin blazing star *(M. pectinata),* it is a branched annual covered with sharp hairs and grows up to a height of about two feet. The leaves are generally longer than those of the above species but have the same unevenly toothed pattern. The flowers of the two species can be distinguished by the upper edge of the petals, which is generally pointed at the center in this species, but notched or flat in San Joaquin blazing star. Occuring on open, rocky slopes in coastal-sage scrub and oak and pine woodlands below 1,200 feet, it can be found in the San Joaquin Valley, the eastern San Francisco Bay Area, and the northern parts of the inner South Coast Ranges. Another more widely distributed species, known simply as blazing star *(M. laevicaulis),* has paler yellow or cream-colored flowers with narrower petals two to three inches long. It occurs throughout most of the sate below 9,000 feet.

Another yellow-flowered plant with the ovary occurring below the sepals and petals is **SUN CUP *(Camissonia ovata),*** a low-growing perennial that often dots hillsides and bluffs near the coast in early spring. In the evening-primrose family (Onagraceae), it has a fleshy taproot and no stem. The four-petaled yellow flowers sit on the ground in the center of a rosette of several oval, often wavy leaves with red edges and veining. The stigma of the ovary is ball shaped, and the inferior ovary is hidden below the leaves, often extending into the ground. It is usually found in clay soils in grassy fields below 1,100 feet near the coast from San Luis Obispo County northward to Oregon.

Sun cup

CALIFORNIA SUN CUP ***(Camissonia bistorta)*** is a taller and more localized species in the same genus with a stem about 20 to 30 inches long. It is often decumbent, creeping along the ground for some distance before becoming erect. The slightly toothed, narrow leaves are in a basal rosette as well as along the often peeling stem. The inflorescence curves over at the top, and each of the four-petaled yellow flowers is about half-an-inch long, generally with two red dots at the base of each petal. The fruit pro-

California sun cup

duced by the inferior ovary is four angled and often wavy or twisted. This sun cup occurs up to 2,000 feet in sandy fields near the coast or in grasslands on clay soils, as well as in openings in chaparral and coastal-sage scrub from Los Angeles and Kern Counties southward to Baja California. It begins to flower in January. More widespread, but smaller, less conspicuous, and with much smaller flowers, is contorted primrose *(C. contorta)*, which occurs in sandy, often disturbed soils in grasslands, chaparral, and woodlands from central California to Washington, Idaho, and Nevada.

Beach evening-primrose

BEACH EVENING-PRIMROSE *(Camissonia cheiranthifolia)* is one of the day-blooming plants in the evening-primrose family. It is a prostrate perennial of sandy beaches with a rosette of gray, densely hairy leaves and several long stems that spread out along the sand. The flowers are less than an inch long, and the four yellow petals, often with two red dots at the base, fade red as they age. The pistil has a round, undivided stigma. It is found from Oregon to Baja California, growing more woody in the southern part of its range. It is in bloom much of the year.

Evening-
primrose

The true **EVENING-PRIMROSE *(Oenothera elata* subsp. *hookeri)*** is another plant found near the coast, but it can occur inland as well. Generally tall and robust, it can grow up to a height of almost eight feet but is generally between two and four feet tall. It is a very hairy, glandular plant with long, more or less lanceolate leaves, generally toothed. The flowers are usually between one and two inches long with four large yellow petals and a long inferior ovary that can easily be mistaken for part of the stem. The sepals are usually reddish, and the hairs have red, blisterlike bases. The flowers open about dusk and wilt the next day when the sun becomes hot. The anthers are more than half-an-inch long. It is found below 500 feet in moist habitats from central California southward. Further inland is found *O. elata* subsp. *hirsutissima*, with the same common

English plantain

name and also occurring in moist habitats. It has green sepals, hairs with normal bases, and anthers less than half-an-inch long. It is found throughout the state up to 9,000 feet.

A common lawn weed in most of the United States is **ENGLISH PLANTAIN** ***(Plantago lanceolata),*** in the plantain family (Plantaginaceae), with a basal rosette of long narrow leaves and a dense spike of many small flowers on an erect, leafless stem. At first glance, the leaves are parallel veined, but a closer look reveals smaller networks of veins as well. The numerous small flowers are generally in four different stages of flowering. The youngest are at the top and still closed, each consisting of a single small white pistil sticking out of an erect, dark, more or less triangular bract. Below these are older, but still closed, showier flowers with four long white stamens with large white anthers

California plantain

protruding from each bract. Next are the open, but less conspicuous, flowers with four small spreading translucent petals that conceal the bract below. The stamens and pistil are gone at this stage, and only the small ovary remains in the center of the four petals. Finally, at the bottom of the spike, is the pointed fruit capsule partly enclosed by the erect bract and containing one to two seeds. This plant was introduced from Europe and is ubiquitous in disturbed places up to about 5,500 feet throughout the state. In wetter areas, common plantain *(P. major),* also from Europe, can be found with broader, rounder leaves and fruit capsules with five to 16 seeds.

A native member of the plantain family is **CALIFORNIA PLANTAIN *(Plantago erecta),*** a much smaller, less conspicuous annual with very thin linear to threadlike leaves one to five inches long. The flowering spike is much shorter than the above species, and the few flowers have four similar spreading petals but much less obvious stamens and pistils. The top of the seed capsule lifts off like a lid to reveal the two seeds inside. Although unimpressive when viewed with the naked eye, the tiny flowers become a work of art when viewed under a hand lens or magnifying glass. This species is found below 2,500 feet throughout the state in less disturbed places, usually on sandy, clay or serpentine soils in grassy areas and open woodlands. The tiny flowers are present between March and May.

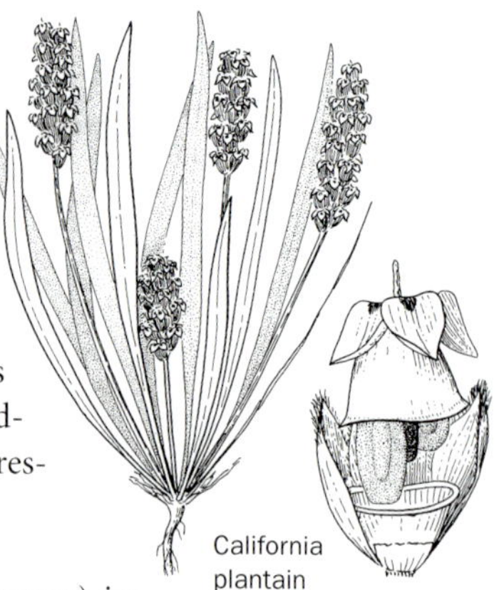
California plantain

In the gentian family (Gentianaceae), is a tall, coarse, glabrous (nonhairy) plant known as **GREEN GENTIAN** ***(Swertia parryi)***, named for Dr. C.C. Parry, who collected specimens in southern California in 1876. Growing to be up to about four feet tall, this plant has leaves edged in white that are narrow near the base of the stem but wider and more oval further up the stem. The petals of the greenish white flowers have purple spots and fringed U-shaped glands. It is a southern species found on open brushy slopes in dry places below 7,000 feet from Los Angeles and San Bernardino Counties to Baja California. Other species occur farther north.

SCARLET PIMPERNEL, or **POOR-MAN'S WEATHERGLASS,** ***(Anagallis arvensis)*** is one of our most common and persistent garden weeds but is very colorful. Introduced from Europe and a member of the primrose family (Primulaceae), it is an annual, almost prostrate plant with opposite leaves and salmon to orange flowers about the size of a dime, although it is occasionally found with blue or white flowers. A close look at the flat

Green gentian

Scarlet pimpernel, or poor-man's weatherglass

Large-flowered collomia, or wild bouvardia

pinwheeled flower reveals an artist's pallette of colors with its five orange petals with burgundy bases, bright yellow stamens, and green pistil. The flowers are very light sensitive and only open on sunny days. It is found in disturbed places throughout California, generally below 3,500 feet.

LARGE-FLOWERED COLLOMIA, or **WILD BOUVARDIA**, ***(Collomia grandiflora)*** belongs to the phlox family (Polemoniaceae). It is an annual with erect stems a foot or more high and entire lanceolate to linear leaves somewhat glandular on the undersides. The flowers are about an inch long and in tight terminal clusters. The color ranges from yellow to salmon to orange or almost white, and the stamens have anthers with blue pollen. It can be found in open areas between 1,500 and 8,000 feet in much of California and blooms in April and May.

Common fiddleneck, or rancher's fireweed

COMMON FIDDLENECK, or **RANCHER'S FIREWEED,** ***(Amsinckia menziesii* var. *intermedia*),** with its coiled branches of flowers and its ovary of four one-seeded nutlets, is a typical member of the borage family (Boraginaceae). It is usually a rough-hairy plant, although sometimes softer, ranging from eight inches to four feet tall and generally branched. The yellow to yellow orange flowers are tubular, spreading into five equal lobes that often have five red orange spots near the mouth of the tube. The flowers are less than half-an-inch long and wide, and the rough fruit is tubercled or ridged. It is found in open, often disturbed areas below 5,500 feet throughout California.

A smaller fiddleneck found near the coast is **SEASIDE AMSINCKIA,** or **SEASIDE FIDDLENECK,** ***(Amsinckia spectabilis)*.** Very similar to the above species, the flowers and nutlets, as well as the plant itself, are generally smaller. Two or three of the five sepals are often partially fused together. The leaves are finely toothed along the edges, helping to distinguish it from other similar-looking species of fiddle-

Bush monkeyflower

neck. It is found on coastal or inland sand dunes from British Columbia to Baja California.

California has many species of monkeyflower *(Mimulus)*, and they come in a variety of colors. In the figwort family (Scrophulariaceae), they are characterized by their angled calyx (fused sepals) and two-lipped corolla (fused petals). One of the more common and conspicuous woody, shrubby species is **BUSH MONKEYFLOWER *(Mimulus aurantiacus)***. Often growing up to six feet tall, it has opposite dark green linear to oval leaves, glandular and sticky on the undersides. The tubular two-lipped flowers are one to three inches long and generally orange, but they can range from white to yellow to red. Common on dry slopes below 5,000 feet throughout California, it is a complex and variable species, hybridizing freely, and many local forms have been identified.

Another common and highly variable monkeyflower, but shorter and herbaceous, is **COMMON LARGE MONKEYFLOWER**

Common large monkeyflower

(Mimulus guttatus). It is a perennial herb with more or less hollow stems and pairs of broad, toothed leaves. The two-lipped yellow flowers, usually spotted red, range from about half-an-inch to almost two inches long with stems exceeding the length of the calyx. In fruit, the angled calyx becomes quite swollen. It has many forms and may grow to a height of two or three feet. It is found in wet places through most of California below 10,000 feet and flowers from March to August.

In this same family is **JOHNNY-TUCK**, or **BUTTER-AND-EGGS**, ***(Triphysaria eriantha)***, an annual growing up to about a foot tall with clusters of pale yellow flowers about half-an-inch long. It belongs to the owl's-clover group, which includes *Triphysaria, Orthocarpus,* and some members of *Castilleja.* This group is

Johnny-tuck, or butter-and-eggs

characterized by two-lipped tubular flowers, with an upper lip forming a beak or hood and the lower lip consisting of three pouchlike lobes. In this species, the beaked upper lip is deep purple, and the pouched lower lip is yellow. It is generally hairy and branched with purplish leaves divided into three to seven lobes. It often grows in great masses and can be found in open grassy places below 4,000 feet throughout California.

The carrot family (Apiaceae) is known in the vegetable garden for parsnip *(Pastinaca sativa)*, celery *(Apium graveolens)*, carrot *(Daucus carota)*, parsely *(Petroselinum crispum)*, and dill *(Anethum graveolens)*. It has small flowers arranged in large compound umbels (segments radiating from

Southern tauschia

one point) and oil tubes with characteristic essential oils, odors, and flavors. A yellow-flowered native species in California is **SOUTHERN TAUSCHIA** ***(Tauschia arguta),*** a glabrous perennial growing from one to more than two feet tall with leathery leaves divided into several oblong to oval, sharply toothed leaflets one to three inches long. The flower stalk is seven to 18 inches long with an umbel of several unevenly stemmed smaller umbels of yellow flowers. It occurs in chaparral and woodland below 5,000 feet in southwestern California and into Baja California. Further north is Kellogg's tauschia *(T. kelloggii)*, with nonleathery leaves divided into more deeply lobed leaflets. It ranges from central California to southern Oregon.

Several species of lomatium *(Lomatium)* occur in California, but the most common and widespread is **COMMON LOMATIUM**, or **BLADDER-PARSNIP**, ***(Lomatium utriculatum).*** A perennial growing from a long narrow taproot, it has multidissected, fernlike leaves. The clusters of small yellow flowers are surrounded by

Common lomatium, or bladder-parsnip

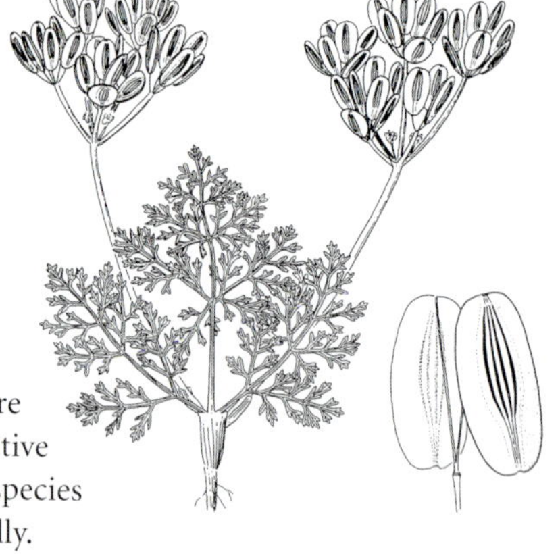

several wide oval bracts, looking like a charming little nosegay of flowers. It can be found on open grassy slopes, in meadows, and in woodlands below 5,500 feet throughout California. The roots of many species of lomatium were used as food by the Native Americans, and some species were also used medicinally.

COAST SILK TASSEL, or **QUININE BUSH**, ***(Garrya elliptica)*** is a very different-looking plant and does not strictly qualify as a wildflower, but its long tassels, or catkins, of flowers are so

Coast silk tassel, or quinine bush

striking that it is included here. It belongs to the only genus in California in the silk tassel family (Garryaceae), but there are several species in the genus. Coast silk tassel is a shrub or small tree with pairs of thick wavy leaves densely hairy on the undersides. The grayish green flowers are petal-less and arranged in long, hanging clusters, or catkins, with the male and female flowers on separate plants. The male flowers are especially lovely, forming, long, narrow silky tassels up to eight inches long. The female catkins are shorter and wider. The tassels first appear in early to middle winter but often last into the spring months. It can be found on coastal bluffs, sand dunes, chaparral, and foothill-pine woodlands below 3,000 feet in most of California and western Oregon.

Twinberry

TWINBERRY ***(Lonicera involucrata)*** is a colorful member of the honeysuckle family (Caprifoliaceae) that is found in wet areas. It is an erect shrub with glandular, sparsely hairy oval leaves one to five inches long situated opposite each other on the branches. The narrow yellow to orange tubular flowers are in pairs with two sets of bracts below. The outer two bracts are leaflike and pointed, but the two inner are more rounded and glandular. When the plant is in fruit, the inner bracts become enlarged and turn deep red or purple, forming an attractive collar below the two black round berries that develop from the inferior ovaries of the flowers. Twinberry is found in wet areas, especially near streams, below 9,000 feet from central California to Alaska.

PEAK RUSH-ROSE ***(Helianthemum scoparium),*** in the rock-rose family (Cistaceae), is a low, bushy perennial with many ascending to spreading stems up to about a foot long with short

Peak rush-rose

narrow leaves. The yellow flowers are about half-an-inch wide with five rounded petals and up to 45 stamens. A distinctive characteristic of this genus is that two of the five sepals are much narrower than the other three. In this species the outer two are more or less linear, and the inner three are wide and pointed. It occurs in dry, sandy or rocky areas in chaparral and other habitats below 5,000 feet in much of California, blooming from March to early summer.

The sunflower family (Asteraceae) has a great many yellow-flowered plants. It is distinguished by its heads of flowers that appear to be a single flower but in reality are clusters of many small flowers packed tightly together in a head. The outer ray flowers are often long, thin, and petal-like; the inner disk flowers are usually small and tubular. Two species in this family, both in the same genus, have the common name of **CALIFORNIA BUTTERWEED.** The first is *Senecio aronicoides,* a perennial herb one to three feet tall with several long-stemmed oval leaves near the base of the plant and smaller leaves along the

California butterweed
(S. aronicoides)

California butterweed
(S. californicus)

stem. The flowering stalk is topped by 10 to 30 flower heads composed of small tubular yellow disk flowers and, occasionally, one or two outer petal-like ray flowers. It is found in dry, open woodlands and forests below 8,000 feet from central California northward to southern Oregon.

The other plant named **CALIFORNIA BUTTERWEED *(Senecio californicus)*** is smaller than the above species and an annual growing from four to 16 inches tall with arching branches and thin, sometimes fleshy, linear to oval leaves one to three inches long. The three to 10 flower heads are about an inch wide each and are composed of many tubular disk flowers surrounded by 13 spreading petal-like ray flowers. Several black-tipped leaflike bracts, or phyllaries, sit below the flower heads. It occurs in dry, open places below 4,000 feet from Monterey and Tulare

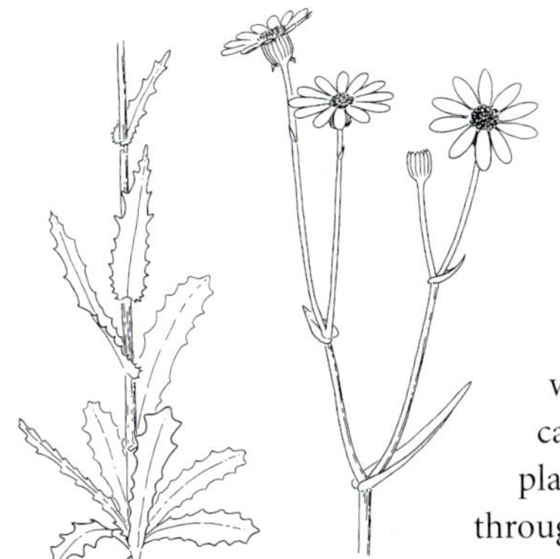

Counties southward to Baja California. More widespread, smaller and very weedy is common groundsel *(S. vulgaris)*, with only the lower phyllaries black tipped, and without any ray flowers. It can be found in disturbed places below 5,000 feet throughout the state.

Also in the sunflower family is **BIGELOW'S SNEEZEWEED**, or **BIGELOW'S SNAKEWEED**, ***(Helenium bigelovii)***, a perennial with stems from one to four feet tall with decurrent leaves, that is, the base of the leaf continues to grow down the stem forming a winglike appendage along the side of the stem. The flower heads have 14 to 20 yellow ray flowers surrounding a rounded central clump of disk flowers that may be yellow, red, brown, or purple. It occurs in wet places, especially along streams,

Bigelow's sneezeweed, or Bigelow's snakeweed

Narrow-leaved mules ears

below 10,000 feet in most of California and into Oregon. It flowers from June to August.

NARROW-LEAVED MULES EARS ***(Wyethia angustifolia)*** is a perennial with long, lanceolate leaves and generally a single, sunflower-like flower head on a stem eight to 12 inches long. The yellow ray flowers are very petal-like, surrounding numerous yellow tubular disk flowers in the center. The leaflike phyllaries below the flower heads are soft hairy. It occurs in grasslands below 6,500 feet in the outer South Coast Ranges, the Sierra Nevada, the San Francisco Bay Area, and from northwestern California to Washington. Other species of mules ears have wider leaves, such as gray mules ears *(W. helenioides),* another common species, but with wide, gray, soft-hairy leaves. It occurs in the Sierra Nevada and from central California northward.

GIANT COREOPSIS ***(Coreopsis gigantea)*** is a stout often unbranched shrub two to several feet high, which during the dry

Giant coreopsis

season greatly resembles a dead broomstick. When the winter rains come, tufts of bright green, finely divided leaves appear at the tips of the branches and are soon followed by many large yellow daisylike heads of flowers. The 10 to 16 ray flowers are about an inch long and surround numerous small yellow disk flowers. It is found on rocky sea cliffs, exposed sand dunes, and shrubby hillsides from Los Angeles to San Luis Obispo Counties and on the Channel Islands. It flowers from March to May.

BUSH-SUNFLOWER ***(Encelia californica)*** forms broad, rounded clumps two to four feet tall with scattered oval to diamond-shaped leaves one to two-and-a-half inches long. The showy yellow flower heads are two to three inches in diameter and consist of outer yellow ray flowers and central purplish brown disk flowers. It occurs on coastal bluffs and inland canyons up to 2,000 feet from Santa Barbara County to Baja California, blooming from February to June. Handling of this plant can sometimes cause an allergic reaction in some people.

Bush-
sunflower

GOLDEN-YARROW ***(Eriophyllum confertiflorum)*** is a perennial herb or subshrub, generally one to two feet high, and somewhat white woolly on the stems and the undersides of the deeply lobed leaves. The yellow flower heads are in flat-topped clusters consisting of 10 to 75 disk flowers and only four to six wide, round ray flowers. It is common on dry, brushy slopes below 10,000 feet from Mendocino, Tehama, and Calaveras Counties southward to Baja California. It begins to flower in April.

SUNSHINE, or **GOLDFIELDS**, ***(Lasthenia californica)*** is a small somewhat hairy annual with narrow, unlobed leaves. Usually only a few inches tall, it has bright yellow flower heads made

Golden-
yarrow

up of six to 13 petal-like ray flowers surrounding numerous small disk flowers. The four to 13 phyllaries below the flower head overlap but are not connected as they are in some other similar-looking species. It grows in very moist, usually flat habitats below 5,000 feet, often forming great masses of yellow that can be seen from a great distance. It occurs throughout California and into southwestern Oregon, flowering from March to May. Other more localized species of goldfields are found in vernal pools, a habitat that is becoming increasingly rare and endangered in California. (See photograph, page 250.)

Sunshine, or goldfields

Pacific grindelia, or Pacific gumplant

PACIFIC GRINDELIA, or **PACIFIC GUMPLANT**, ***(Grindelia stricta* var. *platyphylla)*** is a more or less prostrate resinous and sticky perennial with yellow to greenish stems and fleshy oblong to lanceolate leaves with wide, generally toothed, tips. The

Great Valley grindelia

bright yellow flower heads are one to two inches wide with 20 to 60 ray flowers up to an inch long on the outside and a protective gummy resin exuding from the numerous disk flowers in the center. It is confined to coastal salt marshes and sea bluffs along the coast, blooming in late spring and summer. Two other varieties occur along the coast—*G. s.* var. *angustifolia* is found mostly in central California, and *G. s.* var. *stricta* occurs mostly along the northern coast of California.

A common inland species of gumplant is **GREAT VALLEY GRINDELIA *(Grindelia camporum)*,** with the same resinous flower heads as Pacific grindelia but with phyllaries that are strongly reflexed or sometimes even coiled. The lanceolate to oval leaves are about an inch long and stiff but not fleshy as in Pacific grindelia. The flower heads consist of 25 to 39 ray flowers about half-an-inch long and numerous disk flowers. This species is found in grasslands and woodlands up to 5,000 feet in the Central Valley and much of western California.

Tidy-tips

TIDY-TIPS *(Layia platyglossa)* is an annual, glandular, daisylike plant with narrow, rough-hairy leaves, the lower ones generally somewhat lobed but the upper ones entire. The flower heads are composed of five to 18 yellow ray flowers tipped with white and numerous central yellow disk flowers. It can

Desert-dandelion

be found in many habitats below 6,500 feet in the Central Valley and most of western California. It is common in grassy places at low elevations from Mendocino and Butte Counties southward. It flowers from March to June.

DESERT-DANDELION ***(Malacothrix californica)*** is a fragrant annual with a milky sap. It has numerous linearly lobed leaves, hairy at the base, and leafless flower stems that end in dandelion-like heads of many pale yellow strap-shaped ray flowers. There are no disk flowers. The flower heads are surrounded below by several layers of hairy phyllaries. Found in dry, open sandy places below 5,500 feet, it occurs in many different habitats from central California southward. It blooms from March to May.

TREES

Although not strictly wildflowers, many trees have very brilliant and profuse blooms in spring that attract the average hiker and wildflower enthusiast. A few of the more common and widespread of these attractive trees are included here.

PACIFIC MADRONE *(Arbutus menziesii),* related to manzanita (*Arctostaphylos* spp.) and salal *(Gaultheria shalon)* in the heath family (Ericaceae), is a tall, beautiful tree with a broad crown, large shiny, leathery leaves, and orangish, exfoliating bark. The dense clusters of numerous white urn-shaped flowers appear from March to May and are followed by red to orange berries less than half-an-inch in diameter. Found on wooded slopes and in canyons below 5,000 feet, this tree occurs at scattered stations in southern California and is abundant in central and northern California, ranging to British Columbia. A related

Pacific madrone

California buckeye

species from the Mediterranean region and often used in landscaping is the common strawberry tree *(A. unedo)*, with smaller leaves and rough, round, fleshy, edible fruit.

CALIFORNIA BUCKEYE ***(Aesculus californica)*,** in the buckeye family (Hippocastanaceae), is a large shrub or small tree with a broad round top. It is one of our loveliest native trees with palmately compound leaves of five to seven oblanceolate leaflets ranging from two to seven inches long. The crowded, almost foot-high spikes of fragrant flowers attract many bees in spring, although the nectar and pollen contain cyanide, which can be poisonous to honeybees. The trees sheds their leaves very early in the season, starting as early as July, and the bare branches are often quite conspicuous with their large, brown, pear-shaped, hard-shelled fruits. Although each flower spike holds many flowers, only a few fruits are produced. California buckeye grows on dry slopes and in canyons below 4,000 feet from Siskiyou and Shasta Counties to northern Los Angeles County and flowers in May and June. The Native Americans ground the seeds and spread the meal in streams, temporarily paralyzing fish that then floated to the top and were easy to gather. The remaining fish later revived and swam away, no worse for the wear. The buckeye meal was also sometimes eaten as an emergency food when the oak acorn harvest was poor, but, because of the cyanide content, it had to be thoroughly leached in cold, running water over a period of several days to remove the toxins.

Mountain dogwood

MOUNTAIN DOGWOOD *(Cornus nuttallii)* is related to the famous flowering dogwood on the East Coast. The showiest of our western members of the dogwood family (Cornaceae), it is a deciduous shrub or small tree with opposite leaves about two to four inches long. What appears to be a large four-petaled flower is actually a set of four large white bracts (leaflike structures). Close inspection reveals clusters of small, inconspicuous green or white four-petaled flowers sitting in the center of these four large petal-like bracts. Later in the season the flowers produce red berrylike fruits about half-an-inch long. Mountain dogwood is found in brush and woods below 6,500 feet from San Diego County to British Columbia but is generally more common in the Sierra Nevada and northern part of its range, occurring only infrequently in the mountains of San Diego and Los Angeles Counties. Also common near streams and in other wet areas throughout the state is American dogwood *(C. sericea)*, with clusters of smaller flowers and no petal-like bracts.

BLUE ELDERBERRY *(Sambucus mexicana)*, in the honeysuckle family (Caprifoliaceae), is technically a shrub because it has several trunks, but it looks more like a small tree, and very old trees can actually get quite tall. The branches are arched, umbrella style, and the width of the tree is often equal to the height, which ranges from seven to 25 feet. It has distinctive reddish twigs and opposite leaves divided into three to nine large, toothed leaflets. The myriad of small white or cream flowers are in large, flat-topped clusters. The dark blue, almost black, berries have a white powdery coating, giving them a bluish appearance. It occurs mostly on streambanks

Blue elderberry

and in forest and woodland openings but can also be found on more open slopes where moisture is present. It occurs up to 10,000 feet throughout most of California.

Among the woody members of the rose family (Rosaceae) are cherries, plums, and peaches, all of which have a stone fruit (a large central hard seed covered with a fleshy layer). One of the California natives in this group is **ISLAY**, or **HOLLY-LEAFED CHERRY**, ***(Prunus ilicifolia),*** a handsome shrub with stiff, shining evergreen leaves with wavy, spiny-toothed edges. The cluster of many white flowers produces several red to black fruits about half-an-inch wide with thin, sweetish pulp. It is common in shrubby and wooded areas on slopes and in canyons below 5,000 feet from Napa County southward to Baja California.

Islay, or holly-leafed cherry

Utah service-berry

UTAH SERVICE-BERRY ***(Amelanchier utahensis),*** also in the rose family, is a tall, deciduous shrub that can appear treelike. It has gray to red brown bark, and the leaves are dark green and slightly hairy above, but duller green below. The edges are toothed on the upper two-thirds only, whereas the lower third of the leaf has a smooth, entire margin. The flowers are almost an inch wide with strap-shaped petals and pistils with two- to four-parted styles. The round berrylike fruit is purplish black. It can be found in many habitats throughout most of California between 500 and 6,500 feet.

A lovely, magenta-flowered plant in the pea family (Fabaceae) is **WESTERN REDBUD** ***(Cercis occidentalis).*** A small tree, its flowers appear before the leaves, often in profusion. They are about half-an-inch long, and the showy keel petal is larger than the upper banner or side wing petals. The leaves are almost circular and more or less leathery. The flat, oblong seedpods are dark magenta and highly attractive, sometimes staying on the tree well into the winter months. It is found on dry

Western redbud

slopes and in canyons below 4,000 feet in the inner Coast Ranges from Humboldt to Solano Counties and in the Sierran foothills from Shasta to Tulare Counties, with an outlying colony in eastern San Diego County. The Native Americans used the split twigs for basketry.

Tobacco is largely an American plant, but one of the most conspicuous species now found in California is a native of Argentina, probably having arrived here via Mexico. **TREE TOBACCO *(Nicotiana glauca)*** is a tall shrub or small tree in the nightshade family (Solanaceae), with pendulous tubular yellow flowers an inch or more long. The oval leaves are two to eight inches long with a waxy grayish covering. It is a

Western hop-tree

rather weedy plant, yet somewhat attractive, especially to hummingbirds. It occurs in open, disturbed areas below 3,500 feet in most of the state. California has four native species of tobacco, all glandular and viscid, or sticky, and all with white or greenish spreading flowers. They were used by the Native Americans for ceremonial purposes.

A native member of the citrus family (Rutaceae) in California is **WESTERN HOP-TREE *(Ptelea crenulata),*** with the characteristic aromatic gland-dotted leaves of the family. Growing from six to 16 feet in height, it has deciduous leaves composed of three lanceolate to obovate leaflets about one to three inches long. A flat-topped cluster of several small fragrant greenish white flowers produces flat, tan, round fruits with winged edges and two seeds inside, similar to the fruit of the elm tree. The genus name *Ptelea,* in fact, is from the Greek for "elm." Western hop-tree can be found in scrub and wooded habitats in canyons and flats below 2,000 feet in the Sierra Nevada, Cascade Range, northern Coast Ranges, and San Francisco Bay Area.

GLOSSARY

Anther The pollen-bearing portion of the stamen (the male reproductive part of a plant).

Anthesis The time of blooming, during which the flower is open and functional.

Apetalous Without petals.

Appressed Pressed flat against the stem or another part of the plant.

Axillary In an axil (see below).

Axil The angle formed where a leaf branch meets the stem.

Basal Found at or near the base of a plant.

Blade The expanded, generally flat portion of a leaf, petal, or other structure.

Boreal A term referring to plants native to high latitudes of the northern hemisphere or lower latitudes at high elevations; short for *circumboreal.*

Bract A small, leaflike or scalelike structure usually subtending a flower or cone.

Calyx The outer whorl of a flower, usually green, composed of sepals, and surrounding the petals of the corolla.

Capsule A dry, often many-seeded fruit that splits open when ripe along lines of separation or pores.

Carpel A subunit of the pistil; it may be one or many; fused or free.

Caudex The short, sometimes woody, vertical stem of a perennial plant. The plural is *caudices.*

Cauline Belonging to the stem; not basal.

Chaffy Composed of thin, dry, papery scales or bracts.

Cilia Hairs extending outward along a margin or edge.

Cleft Cut or split about halfway.

Compound Composed of two or more similar parts; often used to describe leaves.

Corm A short, thick, underground bulblike stem without scales, as in crocuses or gladiolas.

Corolla Whorl of flower parts (petals) immediately inside or above the calyx.

Crenate Having a scalloped margin with shallow, rounded teeth.

Decompound Divided more than once.

Decumbent Lying flat on the ground but with the tip of the stem or flowers curving upward.

Disk flowers The cluster of tubular, rayless, generally five-lobed flowers, characteristic of the sunflower family.

Entire Undivided, with a smooth, continuous margin.

Epidermal Pertaining to the outermost cell layer of nonwoody plants.

Fascicle A bundle or cluster of leaves, flowers, stems, or other plant parts.

Filament The usually threadlike, anther-bearing stalk of a stamen.

Filiform Thread shaped; may refer to stems, leaves, flowers, or other plant parts.

Fimbriate Fringed with hairs.

Floret A single flower of the grass or sunflower family. Grass florets include the immediately subtending bracts.

Follicle A dry, generally many-seeded fruit that develops from a single pistil, opening when ripe on one side along a single suture.

Funnelform Widening upward from the base more or less gradually.

Glabrous Without hairs, but not necessarily smooth.

Glaucous Covered with a whitish or bluish, waxy or powdery film.

Glochid A barbed hair or bristle.

Glutinous With a gluelike exudation.

Herbaceous Not woody.

Hispid Covered with bristly, stiff hairs; rough to the touch.

Hoary Covered with white down.

Hyaline Colorless, translucent, and transparent.

Inferior ovary An ovary growing below the calyx.

Inflorescence A cluster or other arrangement of flowers and associated structures on a plant.

Involucre A whorl or group of bracts subtending a flower or a group of flowers.

Keel The two lowermost, fused petals of many members of the pea family; also, a prominent dorsal ridge.

Lobe A rounded segment or division on a leaf or petal margin or on another plant organ.

Margin The leaf or petal edge.

Midrib The central vein of a leaf or other organ.

Node The joint of a stem, or the position on an axis or stem from which structures such as a leaf arise.

Nutlet Any small and dry nutlike fruit or seed.

Ovary The ovule-bearing part of a pistil that normally develops into a fruit.

Ovule A structure within the ovary that contains an egg and may become a seed.

Palmate Lobed or veined so that segments radiate from a common point, as fingers from a hand.

Palmately compound Compounded into two parts; usually referring to leaflets.

Palmatifid Very deeply divided into lobes that spread like the fingers from the palm of a hand.

Panicle A branched inflorescence in which the lowermost or inner flowers open before the flowers at the tips.

Parasite An organism that benefits from a physical connection to another living species, often harming the host.

Pedicel The stalk of a single flower or fruit.

Peduncle The stalk of a flower, fruit, or inflorescence.

Pendent Suspended, drooping, or hanging.

Perianth The floral envelope, composed of the calyx and corolla together.

Petal An individual member of the corolla.

Petiole A leaf stalk, connecting leaf blade to stem.

Phyllary An individual bract of an involucre that subtends a flower head in the sunflower family.

Pinnate Featherlike; having similar parts arranged on opposite sides of an axis.

Pinnately compound Made up of pinnate leaflets.

Pinnatifid A pinnately divided leaf in which the divisions are very deep.

Pistil The female organ of a flower, comprising the ovary, style, and stigma.

Pistillate Having fertile pistils but sterile or missing stamens; female.

Pollen The fertilizing, dustlike powder produced by the anther, containing the male gametophyte.

Pubescence A covering of soft hairs or down.

Raceme A simple, unbranched inflorescence of stalked flowers that open from the bottom to the top.

Ray flowers The outer, often three-lobed, showy petal-like flowers at the edge of a flower head, characteristic of the sunflower family.

Reflexed Abruptly bent or curved downward or backward.

Reniform Kidney shaped.

Rhizome An underground, more or less horizontal stem, distinguished from a root by the presence of leaves, buds, nodes, or scales.

Rosette A radiating cluster of leaves generally at or near ground level.

Runner A slender trailing shoot that takes root at the nodes.

Salverform Having a slender tube, with an abrupt spreading throat.

Saprophyte A plant usually lacking chlorophyll that lives on dead organic matter.

Scape A leafless floral axis or peduncle arising from the ground.

Seed The fertilized and ripened ovule; the earliest product of sexual reproduction in plants.

Sepal An individual member of a calyx, usually leaflike.

Sessile Attached directly by the base; without a stalk.

Seta Bristle or rigid, bristlelike body. The plural is *setae.*

Spatulate Spoon shaped; like a spatula; rounded above and gradually narrowing to the base.

Spike A simple unbranched inflorescence of sessile, or nonstalked, flowers.

Spinescent More or less spiny; spine tipped.

Spore The minute, dispersing, reproductive unit of nonseed plants (ferns and fern allies).

Spp. Abbreviation of the plural of *species.*

Stamen The male reproductive structure of a flower with a stalklike filament and pollen-producing anther.

Staminate Having fertile stamens but sterile or missing pistils; male.

Stigma The receptive part of a pistil on which pollen is deposited and germinates, generally terminal and elevated above the ovary.

Stipe The leafstalk of a fern or of a pistil.

Stolon A runner or horizontal shoot that is disposed to root or to give rise to a new plant at its tip or nodes.

Style The stalk that connects an ovary to the stigma.

Subglobose Somewhat spherical or rounded.

Subsaline Somewhat salty but not excessively so.

Subsp. The abbreviation for *subspecies.*

Subtend To occur immediately below and close to another structure; often used to describe the relative positions of bracts, petals, and leaves.

Superior ovary An ovary that is free from and growing above the calyx.

Ternately compound Compounded into three parts, such as a clover leaf.

Truncate Severely attenuated, appearing cut nearly straight across at the base or tip.

Tubercle A small, wartlike protrusion or nodule.

Tuberous Bearing or resembling a short, thick, fleshy underground stem used for food storage.

Umbel An inflorescence, often flat topped, with three or more pedicels that radiate from a common point like the rays of an umbrella, characteristic of the carrot family.

Var. Abbreviation for *variety.*

Viscid Sticky or glutinous.

Whorl A group of three or more similar structures (e.g., leaves, flower parts) radiating from one node.

ART CREDITS

Photographs credited to the California Academy of Sciences Collection are also credited to their individual photographers.

Line Illustrations

Except as noted below, all line illustrations are taken from the first edition and were drawn by either Stephen Tillet or Richard Shaw.

PETER GAEDE 36, 48 (top), 60, 127, 213

JEANNE R. JANISH 44, 147, 156

PAULA NELSON AND BILL NELSON 18, 19

STEPHEN TILLETT 73

Color Photographs

BROTHER ALFRED BROUSSEAU, SAINT MARY'S COLLEGE 37 (bottom), 40, 63, 109, 122, 200, 260

CALIFORNIA ACADEMY OF SCIENCES 131, 133, 135, 147, 180, 189, 220, 243 (top), 257

MICHAEL L. CHARTERS 227 (bottom)

CHRISTOPHER CHRISTIE 59, 73, 94, 141 (bottom), 150, 151, 163 (bottom), 167, 177, 178, 244 (right)

WILLIAM FOLLETTE ii-iii, vi, xii-1, 20-21, 22, 23, 25, 26, 27, 28 (top, bottom), 30, 31, 32, 33, 34, 35, 36, 37 (top), 38, 39, 43, 44, 45 (bottom), 46, 47, 50, 51, 52, 53, 55, 56 (top, bottom), 57, 58, 62, 64, 65, 67, 68, 69, 71, 72, 75, 76, 78, 80, 81, 82, 84-85, 86, 87, 88, 89, 90, 91 (top, bottom), 92, 93, 95, 96, 97, 98 (top, bottom), 100, 101, 102,

103, 104, 105 (top), 106, 110, 111 (left, right), 121, 123 (top, bottom), 124, 125, 126 (bottom), 128, 129 (bottom), 130, 132, 134 (top, bottom), 136, 137, 138, 139, 140, 141 (top), 143, 144, 145 (top, bottom), 146, 148, 152, 153, 154, 155 (top, bottom), 156, 157 (top, bottom), 158, 159, 160-161, 162, 163 (top), 165 (top, bottom), 166, 168, 169, 170, 171, 172, 173 (top, bottom), 174, 179 (top, bottom), 181, 182, 183, 184, 185, 187, 188, 190, 191, 193, 194, 195, 196-917, 198, 199, 201 (top, bottom), 202, 203, 204, 205, 206, 207, 208, 209, 210, 211, 212, 213, 214, 215, 216, 217, 218 (top, bottom), 219, 221, 222, 223, 224, 225, 227 (top), 228, 230, 231, 233 (top, bottom), 234, 235, 236, 237, 238, 240, 241, 242, 244 (left), 245, 246, 247, 249, 250 (top, bottom), 251, 252 (top, bottom), 254-255, 258, 259, 261, 262, 263

JOHN GAME 120

JESSIE M. HARRIS 129 (top), 229

BEATRICE F. HOWITT 133, 180, 243 (top),

STEVE JUNAK 45 (top), 49, 61, 77, 113, 114, 115, 116, 117, 118, 119, 126 (top), 149, 175, 192, 239, 248, 256

STEPHEN LOWENS 29

DR. ROBERT THOMAS AND MARGARET ORR 189, 220

GLADYS LUCILLE SMITH 135

JON MARK STEWART 105 (bottom), 107, 142, 186

CHARLES WEBBER 131, 147, 257

INDEX

Page references in **boldface type** refer to the main discussion of the species.

ABOUT THE AUTHOR AND EDITORS

Philip A. Munz (1892–1974) of the Rancho Santa Ana Botanical Garden was professor of botany at Pomona College, serving as dean for three years. Dianne Lake is rare plant committee cochair and unusual plants coordinator for the California Native Plant Society, East Bay Chapter. Phyllis M. Faber is general editor of the California Natural History Guides.

Series Design:	Barbara Jellow
Design Enhancements:	Beth Hansen
Design Development:	Jane Tenenbaum
Composition:	Impressions Book and Journal Services, Inc.
Cartographers:	Bill Nelson and Paula Nelson
Text:	9.5/12 Minion
Display:	ITC Franklin Gothic Book and Demi
Printer and Binder:	Everbest Printing Company